Urban Ecology for Citizens and Planners

GAIL HANSEN

AND

JOSELI MACEDO

University of Florida Press
Gainesville

26 25 24 23 22 21   6 5 4 3 2 1

Library of Congress Control Number: 2021942322
ISBN 978-1-68340-252-7

University of Florida Press
2046 NE Waldo Road
Suite 2100
Gainesville, FL 32609
http://upress.ufl.edu

UF PRESS

UNIVERSITY
OF FLORIDA

This book is dedicated to all citizens and citizen scientists who enjoy and treasure nature in their towns and cities, and engage in activities to preserve, protect, and enhance urban nature in all its forms.

Contents

Figures

Foreword

There is no doubt that the world is changing rapidly around us. In recent decades we have seen significant advances in—and alterations to—our global social, technological, and environmental contexts. We have seen geopolitical unrest and the rise of counterfactual realities, nationalism, and antiscience rhetoric. And of course, right now, an urgent global medical challenge.

These challenges demand response from those of us in higher education: a reinforcement of our collective commitment to public and societal well-being. In 1862, the Land-Grant College Act (or "Morrill Act") called for the establishment of a system of public colleges for those who lacked access to higher education, predicated on the idea that the well-being of a nation is dependent on an educated citizenry. This is even more true now. It is through education, and concomitant development of critical and reflective thought, that we will collectively address the challenges we currently face and move beyond narrow, parochial, insular, or fallacious thinking.

This book serves as part of that response. In true Land-Grant fashion, it takes knowledge and makes it accessible to the broadest possible range of people. It addresses a critical emerging need. As the world becomes increasingly urban, there is a need to redefine and reimagine what "nature" means to us and for us. Study after study indicates the critical importance of nature and time outdoors for human well-being, but what does that mean in an urban context? What is or should be our connection to the outdoors in an urbanized world? People have drifted from the ecological lived experience that once came with time spent outdoors and in nature. The natural world has become less and less known to people in urban settings. And while there can be mystery and wonder in the unknown, there is affection and commitment for things that are known. Therein lies the power of this book. It makes urban nature knowable and known. The authors bring expertise in landscape architecture, urban planning, and university extension along with commitment to public education. Drawing on examples from experience in cities worldwide, the book provides a wide-ranging overview of ecology in urban areas. It is designed for accessible learning: each chapter addresses the "what" and "so what" of the main topic, while also providing insight about "now what." As an educator who has used this framing in many contexts, it has been my experience that this is a powerful way to foster learning and engagement with the material. But even more essential than clear and consistent chapter organization is the comprehensive nature of the book's topics. Ranging from the social to ecological to physical aspects of city environments, the book is thorough—exhaustive without being exhausting. It provides a critical overview for anyone interested in seeing the big picture of urban nature.

This is a book for practitioners, students, and citizens. It is a response to a need for us all to redefine and reimagine our connection with nature—urban or other. The basic concept of "Nature for People" acknowledges the value and importance of that connection. And by helping to make the unknown known, this book brings us all closer to it.

Teri C. Balser, Ph.D.
Provost and Vice-President (Academic)
University of Calgary

Preface

This book is for readers who want to learn more about nature in their city. The authors discuss urban ecology with an emphasis on the social and cultural influences that shape urban landscapes. The goal is to encourage citizens, planners, and scientists to actively engage in improving the environment and quality of life in their community through citizen science and planning policy that considers the value and impact of nature in urban contexts. Part I focuses on natural components in and around city structures, such as urban hydrology and urban vegetation that create biodiversity in cities. Part II describes the bionetworks of the city, including the significance of green spaces and environmental design to the overall health of cities and people. Part III explores biophilic urbanism, including our historical relationship with nature, the human affinity for nature, and technology, practices, beliefs, and values that shape the function and look of nature in cities. Each chapter is organized in three sections: What, So What, and Now What. "What" describes environmental conditions and current practices, "So What" explains the significance of these conditions and practices, and "Now What" looks to the future with strategies for improving nature and natural functions in cities, including new approaches and technologies. In addition to these three sections, each chapter describes a project, case study, or expert insight that offers an example to illustrate the chapter topic and explores a citizen science program that relates the contribution of citizens toward integrating natural and urban environments.

The authors would like to thank Australia Awards, Endeavour Scholarships and Fellowships Program, for making it possible to initiate this book project. In 2018 we received an Endeavour Executive Fellowship, which funded the opportunity to gather at Curtin University in Perth to conceive and create a project of international significance. Our research on citizen science inspired us to write a book to encourage more citizens to engage in projects to study and protect nature in urban areas. We would also like to thank the School of Design and the Built Environment at Curtin University for providing the local resources needed to research and develop our first draft. Individuals we would like to thank include Georgina Hansen Stevenson for her help with formatting images in most of the chapters, and our editors at University of Florida Press, Meredith Babb, our first editor who launched us on the path to writing the book, and Sian Hunter, our second editor who cheered us on and guided us to the finish line. We would also like to thank the external reviewers for volunteering their time, and numerous staff members at University of Florida Press for their contributions to this project.

Acronyms and Abbreviations

ACSP   Audubon Cooperative Sanctuary Program
ASIWPCA   Association of State and Interstate Water Pollution Control
            Administrators
ASLA   American Society of Landscape Architects
BEPAC   Building Environmental Performance Assessment Criteria
BES   Baltimore Ecosystem Study
BREEAM   Building Research Establishment Environmental Assess-
            ment Method
CAP LTER   Central Arizona Phoenix Long-Term Ecological Research
CACWNY   Clean Air Coalition of Western New York
CERCLA   Comprehensive Environmental Response, Compensation,
            and Liability Act
C-FERST   Community Focused Exposure and Risk Screening Tool
CPO   Climate Program Office
CS   citizen science
CSS   citizen social science
CWF   Canadian Wildlife Federation
DDT   dichloro-diphenyl-trichloroethane
DEC   Department of Environmental Conservation
DOT   U.S. Department of Transportation
EPA   U.S. Environmental Protection Agency
ES   ecosystem services
ESA   Ecological Society of America
FCM   Federation of Canadian Municipalities
GIBMPs   Green Industries Best Management Practices
GLUSEEN   Global Urban Soil Ecological Education Network
HOA   homeowners association
HUD   U.S. Department of Housing and Urban Development
ICLEI   Local Governments for Sustainability

IPCC	Intergovernmental Panel on Climate Change
IPM	integrated pest management
IUCN	International Union for Conservation of Nature
IVM	integrated vegetation management
LCA	life-cycle assessments
LEED	Leadership in Energy and Environmental Design
LEED-ND	Leadership in Energy and Environmental Design—Neighborhood Development
LID	low-impact development
LIUDD	low-impact urban design and development
MEA	Millennium Ecosystem Assessment
MDG	Millennium Development Goals
NatCap	Natural Capital Project
NACWA	National Association of Clean Water Agencies
NGO	nongovernmental organization
NIHHIS	National Integrated Heat Health Information System
NPCF	National Park City Foundation
NRDC	Natural Resources Defense Council
ORD	Office of Research and Development
PG&E	Pacific Gas and Electric Company
RESES	Regional Sustainable Environmental Science
SCBD	Secretariat of the Convention on Biological Diversity
SDG	Sustainable Development Goals
SHC	Sustainable and Healthy Communities
SITES	Sustainable Sites Initiative
SUDS	sustainable urban drainage systems
TEEB	Economics of Ecosystem and Biodiversity Project
UBR	urban biosphere reserve
UHI	urban heat island
UN	United Nations
UNEP	United Nations Environment Programme
UNESCO	United Nations Educational, Scientific, and Cultural Organization
UNU-IAS	United Nations University Institute for Advanced Study of Sustainability
URBIO	International Network Urban Biodiversity and Design
UrBioNet	Urban Biodiversity Research Coordination Network
USGBC	U.S. Green Building Council
WSUD	water sensitive urban design
WWF	World Wildlife Fund

1

The Science of Cities
Urban Ecology

Few people associate nature with cities, and most would admit to little knowledge of urban ecology or to any understanding of natural processes in urban areas. Yet many city dwellers appreciate the trees, parks, rivers, and lakes in their city, professing a personal attachment to green spaces and the sense of place they create (fig. 1.1). People also understand intuitively that good things come from nature; some are life-sustaining, such as clean air and water; others are benefits that make their city more enjoyable, such as cooler temperatures and opportunities for recreation. These ecosystem services associated with nature in cities make them livable, healthy, and beautiful. While cities are human-made and primarily anthropogenic (human) biomes, humans are not the only inhabitants. Many other animal and plant species live in urban areas, and most people can and do express the enjoyment and feelings of care and concern they have for them, acknowledging the emotional and psychological benefits. This concern was expressed by the citizens of Melbourne, Australia, when the city assigned every tree an email address and asked citizens to email any issues with the trees. The unexpected and delightful result was the number of love letters written to trees in the emails. People thanked trees for all they do for them, wished them well, expressed concern for their health, hoped that no harm would ever come to them, and sometimes just simply wanted to say hello and express their love for the trees (LaFrance 2015).

The Coupled Relationship between Humans and Nature

Ecology of urban areas concerns three interconnected systems that create a city—the natural environment, including soils, water, vegetation, wildlife, and climate; the built environment, including roads and buildings; and the social environment, including people and their activities. The

FIGURE 1.1. London, England: a tree for hiding and climbing to "get away from it all" in a downtown London park.

natural world, of course, does not need buildings or roads; but people do, and it is their needs that create the "urban" in urban ecology. People are the dominant element in the ecology of urban areas, which is why recognizing the social environment is important. The social side of urban ecology considers our beliefs and behaviors, how we value nature, our attachment to nature, why we need nature, and how we create and care for natural urban environments. Urban ecology also considers how we design and build human-made natural areas, with a focus on our design choices and the natural elements we use to create parks, yards, gardens, and stormwater features. And finally, urban ecology is also a field of study that looks to the future, advocating environmental solutions to create sustainable cities with equitable ecosystem services for all city dwellers.

The international scientific journal *Urban Ecology* defines urban ecology as "the study of ecosystems that include humans living in cities and urbanizing landscapes. . . . the term, urban ecology, has been used to describe the study of humans in cities, of nature in cities, and the coupled relationship between humans and nature." (Urban Ecology Field Station 2019). As a field of study, urban ecology is about ecology *in, of, and for* cities that considers the impact of humans on the environment. Urban ecology is unique from natural ecology in that humans highly impact ecology by disturbing soil composition, removing and planting vegetation, changing water flows, reducing air quality, and introducing non-native species. In addition to traditional ecological considerations, urban ecology includes social, cultural, economic, and development pressures. As a result, urban ecology is a multidiscipline field that can play a significant role in improving cities at this defining moment when climate change, growing population, migration, cultural shifts, and economic stressors are challenging sustainability and resilience of cities (McPhearson, Pickett et al. 2016, 199). Resilience and sustainability are concepts that describe the ability of cities to improve their economy, adapt to increasing populations, protect their natural resources, and survive extreme weather events while continuing to meet the needs of inhabitants and maintaining livability. To be sustainable a city must be forward looking, balancing economic, ecological, and human well-being today and for the future. The goal to provide for future populations requires an understanding of the importance of maintaining biodiversity and ecosystem processes to ensure ecosystem services and ensure the continued health of nature and people.

City Nature

People may have the perception that cities are devoid of nature, but to rephrase a quote from the book *Life of Pi* by Yann Martel, "you might be amazed at the animals that fall out if you turned a city upside down and shook it." Squirrels and birds, racoons and rats, frogs, dogs and cats (fig. 1.2), even lizards, snakes, and alligators might fall from the city (Martel n.d.). These animals have become city dwellers; many thrive in cities, adapting quite well to unique food sources and habitats for shelter. Other animals are in cities, not by choice, but because of circumstances. P-22 is the official name of a male mountain lion in Los Angeles that has become the poster boy for the plight of animals that have been trapped by urban sprawl. Nicknamed the "Brad Pitt of the cougar world" in the #SaveLA-Cougars campaign, the handsome cougar is isolated by freeways in his conservation park without a chance of finding a mate. Californians have stepped up and taken responsibility with a plan to build the largest freeway wildlife crossing in the world, using mostly private donations (Solly 2019). Animals aren't the only nature to thrive in cities. Hardy little plants, weeds to some, sprout and thrive in sidewalk cracks and building walls. Green spaces such as residential yards, private gardens, and urban parks support an amazing variety of species; edible plants are grown in community gardens, recreation parks provide grassy playfields, and botanical gardens grow unique and beautiful plants (fig 1.3). Street trees help cool the sidewalk, and planted roadways and utility corridors provide passage into and out of the city as seeds and plants hitch a ride on the cars and people who travel the corridors.

The biological, engineered, and social systems all interact to create complex urban systems. However, the social and built systems of humans exert the greatest impact and create the most profound alterations to the biological systems (McPhearson, Pickett et al. 2016, 206). City planning and design, construction and management of green spaces, and environmental behaviors of citizens impact the health of the environment and people. Urban heat islands are examples of how human behaviors and the social, ecological, and technical systems interact to create unhealthy conditions in the city. The urban heat island effect occurs when tall, closely spaced buildings trap reflected heat, increasing city temperatures. Although the entire city is affected, the lack of urban vegetation, especially large trees, to mitigate the heat island is most prevalent in less affluent areas. Here the elderly, poor, and minority populations have fewer resources for coping and are disproportionally affected based on location, resources, and lack of technical means such as air conditioning. This is one example of the inequitable distributions of ecosystem services that could be improved with

FIGURE 1.2. Izmir, Turkey: feral dogs lying on grass seating steps along the Izmir waterfront promenade.

FIGURE 1.3. Perth, Australia: photos of plants are examples of data used for stewardship and management of parks such as the Kings Park and Botanic Garden that features native plants from SW Australia.

better planning and management of nature in cities (McPhearson, Pickett et al. 2016, 206). Managing ecosystems in urban areas is challenging; aside from the strategies needed to keep natural systems healthy and growing in less than ideal conditions, there is the social expectation for visual quality and provision of recreation opportunities. The recognition of human impact on biological systems evolved over several decades from the concepts of ecology *in* cities, to ecology *of* cities, and currently to ecology *for* cities and the science of cities.

Ecology *in* Cities

The study of urban ecology is a relatively new field in ecology studies. Investigations began in the 1970s when a growing awareness of environmental impacts in rapidly expanding cities pushed biological ecologists to study habitats that were familiar to them in the city context, such as remnant forest patches, meadows, and wetlands and streams. Initially, wildland ecological approaches to management were often modified to adapt strategies to urban conservation. Labeled *ecology in cities*, the focus was on wildlife inventories, botanical surveys, successional studies of woodlands, and the functioning and composition of wetlands and forest patches. The goal of studying natural ecosystems in urban areas was to enhance urban design and human health by providing access to nature and alleviating pollution. Cities began to improve park design, urban planning, and conservation activities as a result of research on ecology *in* cities (McPhearson, Pickett et al. 2016, 201; Pickett et al. 2016, 2–3,6,8).

Ecology *of* Cities

Ecology *in* cities was interested primarily in the natural dynamics of ecosystems but increasingly incorporated social structures and preferences in a transition to *ecology of cities*, which views the city itself as an ecosystem. In the late 1990s, with growing awareness of climate change pressures and the need to consider sustainability and resilience in city planning, ecology *of* cities began to include human components, such as built structures, and open space components. Ecology *of* cities has a multidisciplinary, social-ecological approach that extends to built green infrastructure. For example, in addition to studying natural streams, human-made stormwater drainage networks were also investigated (Pickett et al. 2016, 8). Two things happened when ecologists began to study human-dominated habitats: one, there was a need for a new and precise description of human-dominated habitats, and two, there was a need for an enhanced understanding of the effects of social processes on natural areas and nature in cities (fig. 1.4). These realizations required more research collaboration with the social sciences and other fields such as economics, urban planning, and engineering. Eventually a consensus formed around the idea that interactions between social and natural systems are interconnected and extensive. This new approach to ecosystem ecology focuses more on species identification, biodiversity, spatial organization, and ecosystem disturbance. At the same time, the social sciences have also taken a more contemporary approach to spatial issues such as control and allocation of resources and distribution of ecosystem services. Research collaboration has revealed similarities between contemporary social sciences and ecosystem sciences that have facilitated the social-ecological approach, including in the political process where understanding urban functioning can help create equity in environmental benefits and prevent inequitable

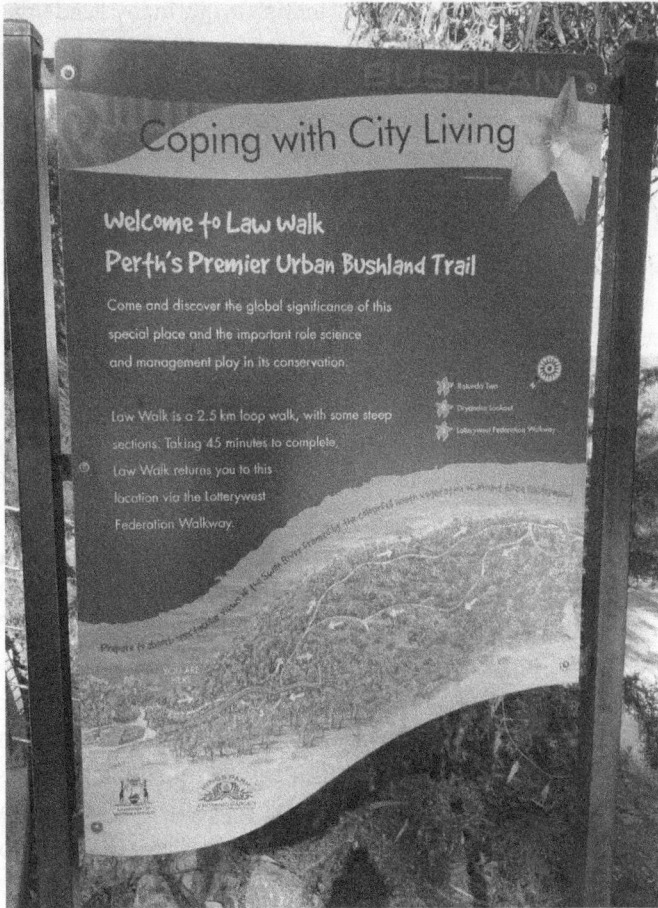

FIGURE 1.4. Perth, Australia: educational sign about an Urban Bushland Trail at Kings Park and Botanic Garden is an opportunity to learn about the effects of social processes on city nature.

impacts from environmental hazards, such as water and soil contamination (McPhearson, Pickett et al. 2016, 202; Pickett et al. 2016, 3–4).

Ecology *for* Cities

The most current urban ecology approach, *ecology for cities*, builds on insights from the first two concepts, *in and of* cities, but links ecological science with civic processes and policy to consider human well-being and urban livability for *applied* urban ecology. Ecology *for* cities is, therefore, transdisciplinary, including urban scientists from many academic disciplines along with city decision makers, policy writers, community associations, and government agencies. Ecology *for* cities includes citizens, planners, and designers in the research process to develop a better understanding of a social-ecological system. This new model of ecology for cities involves researchers in a relationship of shared stewardship; moving from simple bioecological systems to social-ecological systems, and to stewardship systems that improve sustainability and resilience. Stewardship is where urban ecology becomes "knowledge-to-action" or applied urban ecology. Researchers, professional practitioners, urban residents, citizen scientists, and decision makers are included in the dialogue about urban sustainability and resilience, addressing environmental justice, social renewal, conservation, restoration and protection of nature, and the economic strength of cities (McPhearson, Pickett et al. 2016, 203; Pickett et al. 2016, 5,7,8,12).

A new research approach has been put forth that seeks to more clearly address the social-ecological-technical system (SETS) to acknowledge the importance of technology in the urban built infrastructure. Technology is increasingly used to guard against hazards and disasters, many of which are attributed to climate change. The new framework advocates the use of ecologically based infrastructure design in city planning, rather than relying solely on technical engineering for infrastructure design, for example, trading traditional massive flood gates and piped stormwater infrastructure for ecology-based soft structures such as bioswales (fig. 1.5). The goal is to improve sustainability efforts by rethinking infrastructures that are more resilient with collaboration between engineers, designers, architects, planners, and industrial ecologists. A larger research goal is to use local studies, involving local citizens, such as citizen scientists, that consider context, dynamics, and history to generate principles and guidelines for decision makers (McPhearson, Pickett et al. 2016, 207–208, 210).

A New Science Domain

Leading urban ecologists have encouraged a more holistic, integrated science of urban systems—advocating for a new science domain: the *science of cities*. Recognizing that the complexity of cities requires integration of social, economic, ecological, and built infrastructure systems, urban ecologists recommend a diverse research approach that includes working with many disciplines for more knowledge and data sharing, using comparative

FIGURE 1.5. Portland, Oregon: roadside bioswale with planted terraces cleans stormwater as it flows into a nearby created wetland.

research across and between cities, and engaging different information sources, including nonacademic sources such as ordinary citizens, to advance understanding of cities (McPhearson, Pickett et al. 2016, 198).

Emphasizing the Human Dimension

Many books have been written on urban ecology, all with a slightly different focus on the important concepts of ecology in cities. This book emphasizes the human dimension of urban ecology by examining urban ecological concepts in the context of various activities, projects, and programs that are currently being used to create more resilient and livable cities. A primary goal is to encourage partnerships among city officials, researchers, and ordinary citizens by providing information in an easy-to-comprehend format to facilitate collaboration. Decision makers, policy writers, and planners have the opportunity to work with university scientists and citizen scientists to better understand the social-ecological systems in cities and create improved development and growth policies that protect the environment and create the cities we desire.

2

Urban Science Partners

Citizens and Planners

A key concept in this book is understanding how urban ecology science can be applied in planning and policies to help cities make decisions for sustainability and resilience plans. Most information on urban ecology focuses primarily on the biophysical processes of ecology in cities. While this is important for policy, more emphasis is needed on the social and cultural aspects, and the human dimension of urban ecology. Information and knowledge, along with attitudes and values, are the foundations for our politics, policies, organizations, and systems that drive decisions in policy and planning (Douglas and James 2015a, 27). The importance of citizen inclusion has been proven in successful projects around the world. Many examples in these chapters provide solid evidence of better outcomes when citizens participate. This book was intentionally written to be comprehensive in scope, presenting both environmental and social issues, but it is also unique in that urban ecological concepts are illustrated with relatable specific examples, such as the idea of weeds as significant urban vegetation, rather than generic descriptions of ecological and social concepts.

Partnerships for Planning and Policy

Diverse partnerships are essential to city planning and policy. A relatively new partner in the planning process is the ordinary citizen, who is valued for their intimate knowledge of their spot in the city. Citizen groups, neighborhood committees, and local environmental advocates are all examples of ordinary people engaged in urban ecology activities. More recently, a new group of engaged citizens, the citizen scientists, are beginning to have an impact on urban research. Citizen scientists contribute to projects that target cities and build databases for faster data sharing to help quickly advance the science and development of ecology-based cities. Alan Irwin, a sociologist who coined the term "citizen science" in the 1990s, had two definitions: "a form of science developed and enacted by the citizens themselves," and "science which assists the needs and concerns of citizens" (Irwin 2018, 5). The concept that citizens can engage in science to assist their needs and concerns is important; it means they can actively contribute tangible, useful information that decision makers and policy writers rely on to make decisions. An engaged citizenry is key to more informed, and just, planning policies and decisions. On the flip side, it is important for scientists and citizens to understand the planning process, which involves socially driven decisions linked to science, technology, and legal constraints. Scientists and citizens need to understand what information and inputs are needed in planning and the competing interests, including economic, ecological, and social interests, that inform decisions.

More about Citizen Science

Citizen science is a relatively new and dynamic means for "nonexperts" to help researchers and government agencies move the science of cities to a new level of knowledge and understanding. Although there are many other options for citizens to participate in planning and design, such as committees and societies, they vary from city to city and have no central organization. We choose to highlight citizen science because it operates under a well-defined framework that works for any city, within which anybody can participate and contribute to scientific knowledge related to urban ecology. Historically people have been observers of nature for thousands of years, before "scientists" existed, documenting natural phenomena such as annual cherry blossoms in Japan, crop harvests in the United States, and locust outbreaks in many countries, using their knowledge to improve agricultural practices and their community (Dibner and Pandya 2018). The Oxford English Dictionary defines citizen science as "the collection and analysis of data relating to the natural world by members of the general public, typically as part of a collaborative project with professional scientists" (Bonney, Cooper, and Ballard 2019; Citizen Science Central 2019).

A citizen scientist is anyone who volunteers time and effort toward scientific research with professional scientists or who develops projects on their own (fig. 2.1). Common features define citizen science practice. First, everyone can participate, and no qualifications or knowledge of the

FIGURE 2.1. Gainesville, Florida: scientists and volunteers work together to install landscape research plots at the University of Florida.

scientific process is needed. Participants follow a prescribed, widely recognized standard of scientific inquiry, and the data are used by experts and scientists for a specific project or many different projects. However, anyone has access to the data, so scientists and volunteers can work together and share information with the public, or with policy and decision makers. Participants actively engage through data collection and analysis, interpreting results, drawing conclusions, and defining problems, and many personally benefit by doing something they enjoy or improving their community (Dibner and Pandya 2018; SciStarter 2019a).

Citizen scientists have made significant contributions, and data collected by citizens have already contributed to many policy revisions. An amazing variety of projects have been launched by scientists and organizations. The SciStarter website, which is dedicated to citizen science projects, lists more than a thousand current projects. Citizen science projects can happen in places that people use every day, including your backyard. A small sample of projects include collecting bugs in your backyard pool skimmer to analyze biodiversity levels in yards, photographing roadkill to help researchers devise new ways to protect animals, and recording the

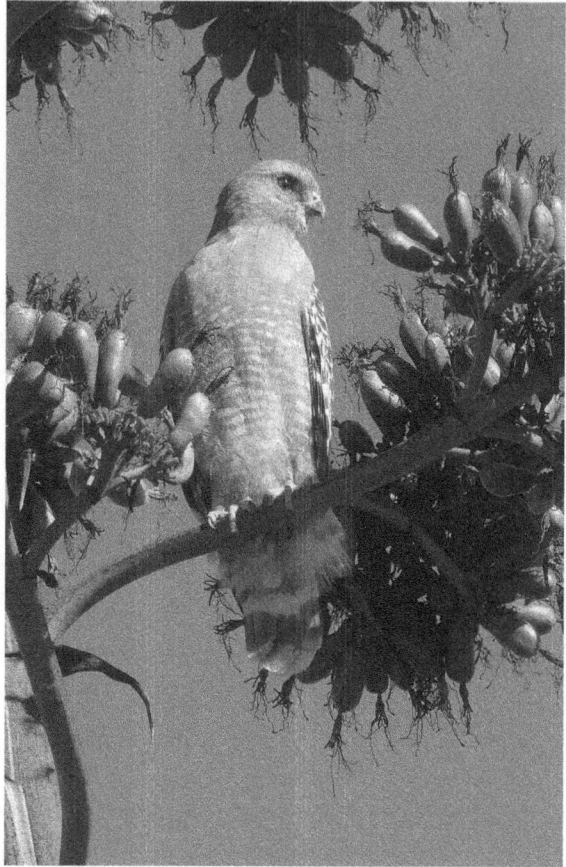

FIGURE 2.2. Monterey, California: sightings of hawks and other birds of prey in urban parks are photographed for citizen science projects such as *eBird* and *Celebrate Urban Birds*.

location of street floods to help scientists prepare for sea level rise and develop new ways to prevent flooding during king tides. Bicycle commuters can attach a sensor to their bicycle to record air temperature and collect pollution particulates to help scientists improve air quality. Hundreds of projects have been launched for bird-watching (fig. 2.2), plant and insect identification, water quality testing in streams and lakes, frog and fish counts, and butterfly and pollinator counts (SciStarter 2019a).

Goals that are frequently cited for participants in citizen science projects include having more citizens gain scientific knowledge and skills so they can participate in community planning and policy development, and encouraging citizens to adopt pro-environmental behaviors that will result in healthier communities. Goals for the practice of citizen science include cultivating a citizenship that supports science. Citizen science also seeks to engage more members of society who represent underserved populations and local people with different local knowledge and perspectives (Bonney, Cooper, and Ballard 2019; SciStarter 2019a).

Citizen *Social* Science (CSS) is a proposed new platform that applies more diverse forms of social knowing and the values and cultures of citizens to policy development. The CSS framework makes citizens co-producers or co-learners in the research process by using the principles of "two-eyed seeing," an approach that blends scientific and indigenous knowledge in which citizens take the lead in making decisions on best policy formation, often over policy makers or the government. The concept is to bring together different ways of knowing. Social knowing and values can help citizens initiate action and respond to policy outside of formal government structures, which helps avoid cultural issues that sometimes emerge from top-down governance and ideology. The key to CSS is having citizen participants with real knowledge to contribute to the policy area, for example, intimate knowledge of the environment and nature based on living and working in the area affected by the policy. In addition to life-long experience, indigenous traditional knowledge is also a valuable resource that citizen social scientists can contribute (Kythreotis et al. 2019, 6, 8).

Decision Makers, Policy Writers, and Planners

Nonacademic participants (those who are not ecology scientists) also include the decision makers, policy writers, and planners who actively engage in solving urban problems in their daily work. One goal of this book is to present ecological concepts that are useful to planning and policy in an applicable format with examples from other cities. Part I describes the natural components and ecological systems to help planners understand how nature functions in cities, and the importance of protecting natural systems in development decisions. Part II describes how nature and the built environment can coexist and function as a system that benefits both people and nature. Part III focuses on the social aspects: the relationship between humans and nature, the benefits we receive, and how information can support policy decisions and a vision for more livable communities.

The chapters follow a similar format for clarity and to help the reader understand the application of urban ecology science for solving environmental problems. Three sections in each chapter present the "What, So What, and Now What" of the topic. "What" describes characteristics, properties, functions, situations, definitions, and key concepts as they apply to the topic. "So What" describes the impacts of human activities as they relate to the topic and why they are of concern for sustainable cities, and "Now What" describes actions, research, and innovations used to mitigate the "So What" impacts. Each chapter also includes a citizen science project and a section on expert insight, case studies, or projects from other cities as examples for planners and decision makers.

PART I

ECOLOGY *IN* CITIES

Ecosystems

Chapters 3 through 9 of Part I describe the natural components, including soil, water, vegetation, wildlife, and climate, that together make up the biodiversity of the natural biosphere and provide ecosystem services in cities. These natural components are the support system of cities, making the built environment and habitation by people possible. A key consideration for optimal function of nature in cities is biodiversity. The health of an ecosystem, especially in the man-made landscapes of cities, depends on biodiversity for peak performance. In addition to the abundance and variety of organisms, the functional traits (functional trait ecology) of plants and animals are important.

To understand how each component contributes to an interconnected biosphere or urban ecosystem, it's helpful to identify how each "part" operates or functions in the built environment of a city, where human activities fundamentally modify natural function and structure. Connecting our health and well-being to the health and functionality of urban soil is one way to illustrate the complexity and importance of nature in cities. For example, soil is normally porous in natural areas, with tiny spaces that allow air and water to circulate to support vegetation. However, urban soils are significantly altered by human activities through disturbance and compaction, preventing water and air from circulating and limiting or preventing plant growth. When plants won't grow, the lack of vegetation reduces wildlife habitat, contributes to poor air quality, and decreases the cooling effect of shade from trees. Lack of shade increases summer temperatures, creating a heat island which impacts the livability of cities and diminishes human health and well-being—all for the want of a tiny air pocket in soil.

In thinking about nature in cities, several ecological concepts have been used to describe the connection between people, nature, and the built environment. To keep our thinking straight, it is helpful to follow the advancement of the study of ecology as it applies to cities to understand how ecology is associated with built environment. Ecology is a basic field of study from which practical applications are derived, such as conservation biology, natural resource management, wetland management, and, more recently, city planning (urban ecology) and human social interaction (human ecology). Ecology includes the study of interactions between organisms and their environment; more specifically, it includes the processes that control the abundance and distribution of organisms, the exchanges among and between organisms, and the transformation and flow of energy, matter, and information among organisms. Ecologists study several biological levels of organization, including individual organisms, populations of organisms, populations in an area called communities, and the interactions of several communities as ecosystems. In other words, ecosystems are a collection of individual plants, animals, and microorganisms (biotic organisms) that aggregate into populations which then form communities that interact with soil, water, and air (the abiotic environment) to create a functional unit whose processes sustain and regulate the environment. All creatures, including humans, live in some type of ecosystem and influence each other and the environment through their behavior. Ecosystems produce natural capital and sustain life-supporting functions, providing benefits to humans in the form of ecosystem services (Boundless Biology–Lumen Learning n.d.; Douglas and James 2015a, 10; Warner and DeCosse 2009d).

At the smallest scale an urban ecosystem is any separate ecological system, such as a park, located within a city, and at the broadest scale the entire city can be considered an ecosystem. In this description ecosystems in cities are sometimes studied as ecosystems within an ecosystem, with each ecosystem being identified by size and physical characteristics. The entire city as an ecosystem can be considered a socioecological system that contains a biogeophysical unit (both the built and natural environments of the city), and the social unit, the individual people, and their institutions such as governance and legal bodies that impact the system (Douglas and James 2015e, 56, 68).

3

Urban Soils

Soil is the ecological foundation of the urban bionetwork. As a living organism it provides ecosystem services such as water filtration, carbon sequestration, and drought and flood mitigation. Soil also provides important wildlife habitat and helps regulate climate extremes. It is key to the health of green spaces in the city and the literal foundation of our buildings and roads, where soil strength is vitally important to our safety and welfare in the built environment. Our health is also directly related to soil health in many ways, including the ability of healthy soil to produce nutrient-rich food and support organisms for creating new drugs. The study of microbiota in soils is on the new frontier of medical research to explore the feasibility of soil fungi and bacteria for new drugs to cure cancer and other diseases. To learn more about soil, the Earth Microbiome Project, a massive crowd-sourced (citizen science) project, is dedicated to mapping and analyzing soil microbial communities around the globe (Thompson 2019).

Although a robust and long history of research on the health of agricultural soils exists, it took growing concern in the face of climate change to fuel a new research agenda on urban soils. New perspectives, and a more urgent need to understand urban soils, have emerged with the onset of large-scale erosion problems and suspected health problems from soil contamination. High soil temperatures have also raised questions about the role of carbon sequestration in soil for climate change adaptation in urban areas (Marcotuillo 2011, 164–174). Municipalities are beginning to recognize that better construction practices and more oversight are needed to protect urban soils, and while geomorphology and geology are well-established fields in soil science, current issues with urban soils will require a new collaborative discipline to address the complex conditions in urban areas. New approaches to protecting and restoring soils must consider engineering and construction technology as a component of ecosystem restoration. Other soil functions related to hydrology, nematology, and microbiology should also be considered in protection strategies and environmental policy. One goal of research is to develop new management strategies that allow for development while keeping healthy soils intact and using new technologies to restore good soil health to previously disturbed soils. Citizen science programs such as the Citizen Science Soil Sampling Protocol (Fibershed n.d.) have traditionally focused on farm and managed land samples to build soil health databases. However, in 2010 the University of Oklahoma Citizen Science Soil Collection Program launched a new project asking volunteers to donate soil samples from their urban backyards for lifesaving drug research (University of Oklahoma 2018). Currently the United States Department of Agriculture, Scientific American (with the University of Oklahoma Natural Products Discovery Group), and National Geographic Society all have citizen science soil collection programs in urban and rural areas to monitor various aspects of soil health, including soil moisture content and microscopic life in soils.

What: Characteristics of Urban Soil

Although all soils are fundamentally the same—a mixture of organic matter, minerals, gases, liquids, and organisms—the basic distinction between natural and urban soil is the quantity of each material, the organization and composition of the mix, and the addition of anthropogenic (human-made) materials. Natural and urban soil often differ in the soil profile, which is the horizontal layering of organic materials and rock (fig. 3.1), and in porosity, the small spaces where water and air circulate (also known as bulk density). Differences also exist in the biological characteristics; including quantity and type of soil biota (organisms), and quantity of mineral and organic matter (chemistry). Healthy soil depends on long cyclical processes at the micro and macro scale to form horizontal layers with a porous structure that allows water and air circulation to support life.

In urban areas construction activities, sealed surfaces, and constant changes in surface material (including roads, sidewalks, and turfgrass) along with changes in elevation (such as walls and curbs) destroy the horizontal continuity and porosity of the soil. Abrupt modifications, such as large-scale soil removal during construction, disrupt the cycle, destroying the layers and composition, compromising the ability of soil to function normally. The change is usually permanent, unless a long-term restoration plan is implemented to help the soil regenerate and reorganize naturally. The transformation from natural soil to anthropogenic soil (a term for the

FIGURE 3.1. Central Italy: soil profile layers visible in a roadside bank cut show the rocky topsoil typical in areas of Italy.

topsoil replaced back after native soil is removed), starts with the scraping and replacing of topsoil. The process creates a mix of materials that often includes on-site topsoil, off-site fill dirt, old building materials, and construction debris (fig. 3.2). The result is compromised soil with limited ecological function to support healthy green spaces in cities.

The quality of soil ecosystem services depends on organic matter that provides carbon and energy for organisms. Globally, soil carbon pools are three times greater than the carbon stored in all land plants (Marcotuillo 2011, 179); however, not enough is known of the net effect of urban activities on the carbon cycle. Recent research suggests that urban activities can both increase and decrease soil organic carbon pools. After an initial loss with soil disturbance, carbon pools appear to rebound in the most highly managed soils. The key seems to be the type of land use and management; for example, golf course fairways can increase organic matter after planting turfgrass. A comparison of carbon stocks in residential turfgrass and native soils showed that the management of turfgrass tends to homogenize carbon densities in cities, despite different native soils, topography, and climate (Marcotuillo 2011, 180). Climate change also presents challenges for urban soil. Most places are predicted to have hotter drier summers and warmer wetter winters, with extreme weather such as heat waves, drought,

FIGURE 3.2. Gainesville, Florida: soil contaminated by debris on construction site. Debris including wood, metals, adhesives, grout, paint, and concrete slurry are commonly left in soil after construction.

flooding, and sea level rise, which could all degrade soils. More research and soil threat models are needed to create policies, adapt protection strategies, and build soil resilience (Defra 2009).

So What: Ecosystem Threats and Human Health Concerns

Threats to soils in urban areas include erosion, compaction and sealing, contamination, and loss of organic matter and biodiversity. Annual costs to society for soil degradation range from $300 million for salinization to $14 billion for erosion (Panagos, Montanarella, and Jones 2006). Soil erosion is typically the slow wearing away of topsoil, either by wind or by water; however, scraping the topsoil on a construction site causes quick and damaging erosion. Urban soil is particularly susceptible to erosion, because properties of urban soil such as poor drainage, poor structure, and salinization can accelerate the erosion process (Ritter 2018). In addition to erosion, the soil properties most highly modified by humans are compaction, nutrients, temperature, and pH. The ecological consequences of changing physical characteristics such as compaction and surface crusting (soil sealing) include reduced aeration and infiltration, different temperature regimes, reduced microbiomes, and reduced nutrients and fertility (Sauerwein 2011, 51). Compacted soil is the most problematic, behaving much like an impervious surface, with small pore size and surface crusting creating a thin hydrophobic (water repellent) layer on the soil surface, which reduces water holding capacity, the exchange of oxygen and carbon dioxide, and the cycling of nutrients, making it difficult to support plant growth (NRCS USDA 2000; Sauerwein 2011, 52). As a result, mechanical aeration, frequent fertilizer applications, and irrigation are needed to maintain plant material (Hanks and Lewandowski 2003). Reduced plant growth equals less biodiversity in plants (especially native plants), which in turn diminishes the healthy microbe habitat the plants need to grow. The loss of microbes, plant diversity, and native plants in turn reduces the ability of the ecosystem to support animals and insects, especially those that rely on specific native plants for food or nesting.

It is not well known how compacted and sealed urban soils with high pH and pollutant concentrations affect the soil organisms and in turn the biodiversity responsible for ecosystem services. Research suggests fewer species numbers and reduced biomass of soil organisms in urban soil; however, some urban soils are quite high in biodiversity. Soil biota generally fall into three groups: macrobiota (burrowing animals, earthworms, and large arthropods), mesobiota (small arthropods), and microbiota (nematodes, bacteria, fungi, and viruses). These soil organisms protect plants from disease and improve soil structure, biofiltration, and fertility (McDonald, Stenn, and Berger 2011). Earthworms are the most important macrobiota, since their feeding process changes the mineral composition and organic matter in the soil. In some urban-rural studies the average number of earthworms was significantly higher in urban areas (Marcotuillo 2011, 176). Urban arthropods (mesobiota) are the least understood soil fauna, but in general, urbanization has caused a decline in arthropod populations. It is suggested that pollution and changes in hydrology are the main reasons. Certain types of urbanization, however, can actually create habitat for some arthropods. Soil invertebrate diversity can be high in parks and green spaces, especially for soil mites, potworms (enchytreids), springtails, ants, earthworms, and the larvae of beetles, aphids, flies, spiders, and snails (Marcotuillo 2011, 177). Microbiota species that decompose soil are often high in number and diversity in urban soils; for example, there can be hundreds to thousands of bacteria and fungi species in 1 gram of urban soil. However, some studies suggest that microbiota activity decreases with urban density and pollutant loads (Marcotuillo 2011, 178).

Chemicals from landscape maintenance, such as pesticides, fertilizer, and herbicides, often leach into the soil and increase the chemical load; however, contamination of urban soils is not limited to new construction or landscaped sites. Some urban areas also have a legacy of contaminated land. Chemical soil saturation from past industrial sites can be a significant risk to human health and the environment. Heavy metals, solvents, and polycyclic aromatic hydrocarbons (PAHs) are just a few of the pollutants. Environmental Protection Acts now require contaminated land to be identified and remediated. However, remediation techniques themselves can often have severe environmental impacts, creating a need to shift to more sustainable remediation solutions, which in some cases may mean "only when necessary." Brownfield redevelopment incentives and new remediation practices and regulations have helped to ensure that policies are appropriate to protect human health. Recent research suggests that urban soil conditions are becoming more homogeneous as the same management practices and technologies are used with increasing global connectedness (Marcotuillo 2011, 166).

Pollutants in urban soil are also a good example of the impact of human activities on soil composition. The most commonly studied pollutants are heavy metals and PAHs. Heavy metals, both natural and introduced by humans in soil, are toxins that do not decay and persist in the environment. Metals such as arsenic, chromium, copper, mercury, and zinc can be found in fertilizer, pesticides, fungicides, and sewage sludge (Marcotuillo 2011, 170–171). However, it is important to remember that almost all of the heavy metals are also natural, coming from the parent rock material that formed the soil. It is also important to note that higher levels of heavy metals have been found in developing cities with emerging economies, compared with developed world cities where one would expect historic accumulation to create higher concentrations. Concentrations also vary widely within cities depending on the density of industrial activities and local weather conditions. PAHs are released in the burning of fossil fuels (coal and oil), and seven compounds have been classified as probable carcinogens by the U.S. EPA (Environmental Protection Agency). Recent research has identified PAH compounds in the street dust and soil of cities where motor vehicle use is heavy. The highest concentrations are found in the surface soil of street shoulders and interchanges. Direct comparisons with rural soils show higher urban concentrations, and vegetable samples from urban areas show a PAH burden 10 times higher than vegetables grown in rural areas (Marcotuillo 2011, 174).

Now What: A Better Management Approach

Soil is increasingly viewed as a critical living organism in urban areas, and studies are growing as a result of increased realization of the importance of soil in ecosystem function and as a resource for humans beyond food and fiber production. Climate change, sustainability, and the use of urban agriculture to meet future food demands are also driving new research. One issue of general concern is the role of soil carbon stores in climate change mitigation. Protecting the carbon store is a priority in the United Kingdom's new soil strategy, which notes that UK soils store about 10 billion tons of carbon in organic matter, mostly within peat habitats, and losses of this soil carbon would have a major effect on climate change. Objectives include understanding the potential to increase stores, reducing the loss of stored carbon, and predicting carbon fluxes with changes in land use (Defra 2009). Soil protection strategies are currently viewed as best management practices for all urban areas to protect carbon stores and soil health.

Soil protection strategies for construction sites are a relatively new objective for sustainable development, although currently many municipalities have no requirements for restoring healthy soil post-construction. The most common strategies include preplanning for protection zones, better site clearing and stockpile techniques, identifying haul routes and stockpile management areas, and improved topsoil replacement and soil amendment practices (Young and Morrison 2012). The UK has developed a set of strategies, *Construction Code of Practice for the Sustainable Use of Soils on Construction Sites,* that sets out parameters for preconstruction planning, soil management during construction, and preparing soils for landscape and habitat creation (Defra Soil Policies Team 2009). Preconstruction planning recommends a soil resource survey by an experienced soil scientist that is incorporated into a site working strategy, and the use of sustainable drainage systems to promote infiltration beyond the construction phase. Managing soil during construction includes a Soil Resource Plan (SRP) that shows location and type of topsoil and subsoil to be stripped, designated haul routes, and soil stockpile management. It also recommends stripping and replacing soils in dry conditions, using tracked equipment to reduce compaction, and utilizing short soil storage periods with separate stockpiles for different soils. The expected after-use for the soil in each stockpile should also be identified. Recommendations are also given for preparing soil for landscapes, such as bonding or "zipping" the topsoil to the subsoil to ensure the condition of the entire soil profile will promote aeration, drainage, and root growth (Young and Morrison 2012).

It can take 1 to 3 years for soil structure to stabilize and provide proper drainage and aeration for plants, so it is recommended to maintain soil health by de-compacting and aerating new turfgrass and monitoring conditions in plant beds. The UK also has companion legislation, such as the *Good Practice Guide for Handling Soils* (MAFF 2000), that provides advice for operators on soil handling and best machinery for stripping, stockpiling (fig. 3.3), excavating from storage mounds, de-compaction, and cultivation of soil during construction. Other recommendations include using temporary vegetation to hold soil in place and clearing land only as construction progresses (Hanks and Lewandowski 2003). While this guide is intended for use in the UK, the strategies can be adapted to any geographic location and are targeted toward developers, contractors, and regulators.

Waste management during construction and demolition (C&D waste) is critical for soil protection. Different construction projects often generate different waste products requiring flexible strategies to reduce, reuse, recycle, or dispose of waste. Techniques include separating waste streams to increase potential for reuse or recycling. For example, some demolition materials (doors and windows) can be reused, and some waste, such as old concrete, is considered clean fill, which can be also be reused, while others are considered solid waste and can be taken to a resource recovery facility. Controlling inventory and selecting products that will produce less waste is also helpful (Utah Department of Environmental Quality n.d.).

Soil fertility and health scores that look at chemical, biological, and physical indicators have been developed for rural and agricultural soils. Water capacity, surface hardness, organic matter, respiration, active carbon, and pH and phosphorus are just a few of the indicators that are measured. The goal is to establish standards for "current best available" methods to assess urban soils and urban soil health indicators to strengthen the science of soil health, develop management decision tools, and create a citizen science portal (Moebius-Clune et al. 2016).

Management is also being improved through research and data collection on soil in urban habitats. The Global Urban Soil Ecological Education Network (GLUSEEN) is an open-source ecological network to encourage investigation with a worldwide multicity comparison of the effects of urban environments on decomposition and soil community structure. Questions that GLUSEEN addresses include how the development of soil communities differ in urban habitats, the importance of native vs. anthropogenic

FIGURE 3.3. Gainesville, Florida: soil improperly stockpiled on construction site. Soil should be covered and free of weeds.

factors on soil characteristics, and the value of using observations of structure and function of urban soil by citizen scientists to advance understanding of soil ecology. One CS project underway is an international effort to collect plant decomposition and soil improvement data using tea bags. A simple and cheap method was developed by Utrecht University researchers to measure decay rate of plant material using tea bags. Decay rate data are used to learn more about the role of soil in the global carbon cycle and improve modeling for climate change. Participants bury tea bags of green tea, dig them up three months later, measure the weight of the bags, and enter the data in an online portal. The microbial communities in the soil are also analyzed with sequencing analysis (Szlavecz n.d.; Utrecht University 2016).

Citizen Science Program: Soil Collection

The University of Oklahoma (OU), Natural Products Discovery Group, launched a popular Citizen Science Soil Collection program in 2010, asking for volunteers to donate soil samples from their backyard so OU investigators could obtain fungi that make special compounds (known as natural products) used to create lifesaving drugs. Participants dig teaspoons of soil in their backyard and send the samples to OU for testing and fungi extraction (Citizen Science Soil Collection Program 2019).

Expert Insight: Restoring Urban Soil

New studies have shown that it may be possible, with the use of soil amendments, to restore urban anthropogenic soil to viable soil in small-scale areas, such as urban tree pits. Dr. Bryant Scharenbroch, a soil researcher at Morton Arboretum and University of Wisconsin, has done field trials of three different growing media to assess effect on tree performance. The experimental soils included native soil (100% topsoil), tree soil (60% native topsoil, 15% compost, 25% sand), and urban soil (60% sand, 25% native topsoil, 15% compost). Tree growth indicators, including leaf size, tree dimensions, and soil respiration, were monitored for one year. Overall the best mix for trees was the "tree" soils: 60% native topsoil, 15% compost, 25% sand. The findings were encouraging, because it means that 100% replacement of the native topsoil was not necessary. Other recommendations for salvaging dirt on site include not moving existing topsoil off-site, keeping clods, rocks, and wood chunks smaller than 2 inches in diameter in the soil, not moving the topsoil more than three times during construction, avoiding overmixing the sand and compost, and not compressing the soil with anything heavier than a foot when planting (MacDonagh 2016).

4

Urban Hydrology

Urban hydrology is about water in cities. Available clean water has always been a challenge for human settlements; massive water structures in ancient cities, such as the Roman aqueducts in Italy and underground cisterns in Turkey, give testament to the enormous efforts to capture, move, and store water for citizens of early cities. Today, water issues remain the same, but the concerns about clean available water have new significance due to increased threats from water shortages, water pollution, and changing climates in many cities. The concepts of managing water remain much the same, including the capture, movement, storage, infiltration, and use of water. What is new is water management technology, the immense scale of urban areas, larger population size, and increasingly negative environmental impacts. The effects of urbanization on water are numerous; the impermeable surfaces in cities modify the natural movement of water, reducing the amount of infiltration and changing the runoff rates. Nonporous surfaces also change the volume and direction of runoff, and the quality of water, which ultimately impacts the ecology of natural water bodies (Douglas and James 2015d, 133–134).

Urban hydrology is also a field of environmental science for the study of water quality in urban areas as it relates to flood protection, environmental and public health, and livable, aesthetically pleasing communities (Fletcher, Andrieu, and Hamel 2013, 261). The discipline combines water science, environmental policy, and public health policy for the regulation of drinking and irrigation water. Urban hydrology requires specific knowledge of hydrological systems, geomorphology, and engineering for flood control to develop stormwater management plans and design standards for stormwater systems. Currently the trend is toward restoring predevelopment flows and water quality to benefit the environment and improve livability in urban areas (Fletcher, Andrieu, and Hamel 2013, 261). Wetland and stream restoration (fig. 4.1) are included in most management plans

FIGURE 4.1. Gainesville, Florida: a concrete canal built in the 1920s was removed to restore the original streambed with planted aquatic vegetation to clean water flowing into a state preserve.

today, requiring specialized knowledge in ecosystem restoration, aquatic plant material, wildlife habitat, and an understanding of the recreation value and landscape preferences for aquatic habitats. The primary goal of managing urban wetlands and streams is to reduce flooding, maintain clean water, recharge groundwater, and create healthy aquatic habitats.

The social dimensions of managing urban water are also important. Flooding, water shortages, poor water quality, and inadequate stormwater infrastructure create public health and livability issues. Management plans need to consider infrastructure and spatial planning, especially in high-density urban areas prone to flooding and water quality problems. Resistance to change by developers and the lack of legal mandates and development policy have been identified as two major obstacles for more sustainable stormwater management in the United States, especially in cities with lower socioeconomic status where best management practices are difficult to implement (Barbosa, Fernandes, and David 2012, 6792). Several citizen scientist programs are contributing important data, including water quality testing and aquatic plant and animal monitoring, which are used by authorities who make urban water decisions. For example, the

Wisconsin Department of Natural Resources maintains a database, called the *Surface Water Integrated Monitoring System* (SWIM), where professionals and volunteers can add and use water data. New and updated policies have been influenced by the baseline monitoring done by volunteers, including granting impaired designations, upgrading protection strategies, and limiting fishing activities.

What: Components of an Urban Stormwater System

In the natural environment the hydrological system is a network of natural features, including streams, rivers, wetlands, ponds, and lakes, that collect, move, store, and clean rainwater. In these undisturbed areas rainwater flows over porous soil and organic material, and much of it is absorbed, preventing flooding, providing water to vegetation, filtering water, and recharging the aquifer. Unlike natural areas, in the built environment rainwater flows over large areas of impervious surfaces, gathering debris and pollutants and disrupting the critical functions of providing water for plant life, recharging the aquifer, and flood control. City water is moved and managed in several ways, including in modified natural streams and rivers, through constructed stormwater systems (fig. 4.2), and in underground pipes that carry drinking, waste, and irrigation water. The systems connect in several ways: storm and irrigation water flow into human-made and natural water bodies, treated wastewater is discharged into natural water bodies, and potable drinking water is often used for irrigation (Douglas and James 2015d, 133–134). Urbanization creates four primary impacts on the ecology of natural waterways: first, the hard surfaces that increase runoff during rainfall events result in too much water entering waterways in the short term; second, the sealing of surfaces limits infiltration and underground flows to streams, resulting in too little water over the long term; third, water quality is impacted by pollution and litter; and fourth, habitat degradation lessens the ecological value of waterways (Cooper, Crase, and Maybery 2017, 70).

To control water in urban areas, natural systems have been replaced with a built system of gutters, storm drains, pipes, canals, and stormwater ponds, collectively known as a stormwater system. Although the original purpose of stormwater systems was sediment collection and flood and erosion control, new approaches consider stormwater a resource to be managed and used rather than simply collected and stored (Fletcher, Andrieu,

FIGURE 4.2. Gainesville, Florida: a constructed stormwater channel to transport water to detention basins on the University of Florida campus.

and Hamel 2013, 262). Today stormwater is also used as an aesthetic amenity in many communities, and systems are expected to provide for groundwater recharge, pollution control, recreational uses, reuse water, and some natural aquatic ecosystem functions, such as wildlife habitat. However, many aging stormwater systems are no longer adequate to control large volumes of water, and many have become conduits for urban waste, creating polluted, nonfunctioning, and visually unappealing water bodies. As a result, some communities are rethinking their stormwater management strategies, including the use of human-made wetlands designed to replicate, as much as possible, the function of natural systems. Two categories of stormwater technology—infiltration-based and retention-based—are in use today. Infiltration-based includes vegetated swales, stormwater basins (fig. 4.3), rain gardens, and porous pavements that help recharge subsurface flows and groundwater. Retention-based (stormwater harvesting) includes wetlands, ponds, vegetated roofs, and tanks or storage basins that help retain stormwater to reduce outflow, enhance pollutant removal, and provide for irrigation (Barbosa, Fernandes, and David 2012, 1794).

Another component of urban hydrology, landscape irrigation, is also proving problematic. Urban landscapes are typically turf dominated and often have ornamental plants that require supplemental water to survive. The use of water-thirsty plants and a lack of knowledge about best irrigation practices contribute to excess water use and increased water pollution. Surplus water from irrigation and rain events typically sheet-flow over predominately turf landscapes and carry pollutants that eventually end up in stormwater ponds and natural water bodies. The contaminated rain and irrigation water is saturated with urban waste such as fertilizer, herbicides, pesticides, and animal fecal matter that impair water quality, destroy wildlife habitat, and create health problems.

FIGURE 4.3. Portland, Oregon: curb cuts allow water to flow into vegetated stormwater basin for parking lot.

So What: Poor Environmental Conditions and Human Health Concerns

Because the original purpose of engineered stormwater systems was to prevent flooding, there was little consideration for possible social or environmental consequences, with little expectation for the system to provide ecosystem services to animals or humans. Communities have started to recognize the negative impacts and value of urban water, and efforts are being made to understand the environmental and social issues around water quality and use. Urban stream syndrome describes ecological consequences such as loss of sensitive species and organic matter and increases in nutrients and toxins even at low levels of urbanization (Fletcher, Andrieu, and Hamel 2013, 264, 266).

The traditional focus on stormwater quality has been about sediments, nutrients, and heavy metals, but more recent studies have identified a large number of pollutants, such as hormones, synthetic chemicals, and pesticides as an emerging problem (Fletcher, Andrieu, and Hamel 2013, 267). Pathogens and emerging priority pollutants that affect health include herbicides, industrial components, and organic micropollutants, such as PAHs (polycyclic aromatic hydrocarbons, from the burning of coal, oil, gas, and wood) and PCBs (polychlorinated biphenyls, from toxic industrial waste compounds) (Barbosa, Fernandes, and David 2012, 6789; Fletcher, Andrieu, and Hamel 2013, 267). PAHs can be found in sediment of rivers and estuaries where they impact the health of sediment-feeding organisms, and they are classified as probable human carcinogens with links to several cancers. Pollution such as nutrient loading (nitrogen and phosphorus) leads to eutrophication, which is the rampant growth of certain biological organisms such as algae. Large masses of algae reduce the clarity of water, reducing light and limiting plant photosynthesis and oxygen production, leading to a loss of aquatic plants and wildlife habitat. Extreme eutrophication can also cause large algal blooms that produce toxic chemicals and noxious odors when they decompose. Usually attributed to excess nitrogen and phosphorus (typically from fertilizer, septic systems, and atmospheric fallout), nutrient loads often kill fish, small amphibians, and beneficial insects that provide food for wading birds and other aquatic animals (Havens 2018, 1–2).

For stormwater management purposes, multiple sample collections are used to test each pollutant along a range of concentrations (maximum, minimum, and standard deviation) to obtain a rainfall event mean concentration (EMC) of that pollutant. EMC is calculated as the total mass of pollutant divided by the total volume discharged. A site mean/median concentration (SMC) is then calculated, which is the mean or median of all the measured EMCs. Another parameter used to describe stormwater quality is the pollutant annual mass load per unit area. This method typically shows the annual loading rate to be higher in high-density residential areas and decreasing as it goes through low-density residential, industrial, and undeveloped land uses. But it also depends on the type of pollution and type of industries; for example, large quantities of fecal bacteria (from dogs, cats, and birds) have been observed in undeveloped areas, indicating natural sources of contamination as well (Barbosa, Fernandes, and David 2012, 6789).

Ponds also contribute to ideal conditions for undesirable insects such as mosquitoes. Attempts to control mosquitoes often mean the use of larvicides to destroy the larvae or the introduction of exotic species such as Mosquito Fish (*Gambusia holbrooki* and *G. affinis*), which in turn feed on the eggs of native frogs, crustaceans, and fish, depleting the food source for wading birds and resulting in more ecological imbalance. The loss of wading birds and other wildlife that homeowners enjoy can have a negative impact on the desirability of the community and the property value of individual homes. The visual quality of stormwater ponds is also highly valued by homeowners; however, many aquatic plant choices for ponds are made based on aesthetic appeal rather than ecological considerations, which can further disrupt the ecosystem balance. Plants that may help the uptake of nutrients sometimes don't fit the aesthetic preferences of homeowners, who describe them as weedy or "swampy" in appearance. Mechanical means, such as aerators, are sometimes employed to decrease algae; however, they have limited success and often add to the loss of aesthetic appeal. Community maintenance professionals often prefer not to use aquatic plants to simplify maintenance, choosing instead to have turf to the water's edge. Without plants to buffer the constant motion of small waves on the shoreline, the soil below the turf washes away, resulting in severe erosion and collapse of the bank, further affecting the ecology in the pond and reducing the size and value of the property (Hansen and Hu 2013).

Now What: A Better Ecological Approach to Urban Hydrology

Urban hydrology has taken on new meaning in many communities. As our understanding of the ecological and social benefits of urban waters has evolved, the management of stormwater has become a socioenvironmental issue, creating new challenges for policy, funding, and treatment of water. New technology, adaptive strategies, improved ecological knowledge, and a more anthropocentric approach are leading to better design options that benefit communities in multiple ways. Several terms describe these new approaches, including Sustainable Urban Drainage Systems (SUDS), Water Sensitive Urban Design (WSUD), and Low Impact Development (LID). The primary goal of these systems is to maintain predevelopment hydrologic function by using infiltration, detention, and evaporation to control stormwater (Burns et al. 2012, 231; Fletcher, Andrieu, and Hamel 2013, 268). Objectives of these approaches include sustainable watershed management, restoring water quality, conserving water resources, and enhancing urban landscapes (see chapter 13, Urban Structures, for further discussion). Engineered solutions such as green roofs, cisterns, stormwater planters, and subsurface detention have also become more popular treatment options (Fletcher, Andrieu, and Hamel 2013, 268). In highly built, well-established urban areas, other strategies may be employed, often requiring the removal of impervious surfaces, including roads, parking lots, and buildings to create more green space, and bringing streams and rivers that were channeled into pipes underground to the surface to reclaim their original streambed path. Daylighting of streams (fig. 4.4) in many large cities has proven a very successful strategy to control flooding and create green corridors for wildlife habitat and recreation. Walking promenades along the banks of surface streams also improve livability by encouraging exercise, providing contact with nature, and creating economic opportunities along the stream corridor. A similar strategy includes creating green flood plains that provide opportunities for economic stimulus with nature-based, wildlife-focused, recreation opportunities (Hu and Keeley 2013).

Restoration and conservation of natural wetlands is another strategy, as is the design of novel landscapes, such as human-made wetlands, where a network of created wetlands collect stormwater citywide. Although wetlands are more difficult to restore or re-create because of their ecological complexity, novel wetlands are designed to function in a similar fashion, by cleaning water and providing habitat, but do not necessarily look like a

FIGURE 4.4. Portland, Oregon: daylighted creek with newly planted vegetation. The original creek bed was restored during construction of a new housing development.

natural wetland (fig. 4.5). Strategies are also being put in place to protect natural bodies of water, including using vegetative buffers that separate the water from pollution sources, enacting fertilizer ordinances to reduce fertilizer, and using plants that require less fertilizer. A new paradigm has also emerged for an integrated total water-cycle model designed to meet

FIGURE 4.5. Perth, Australia: newly planted floating wetlands clean water at Chevron Parkland.

several objectives, including maintaining public health, preventing flooding, preserving ecosystems, and reducing the effect of urbanization on the water cycle. The challenge is developing models that can simulate all the components in a combined way. Models such as MUSIC (Model for Urban Stormwater Improvement Conceptualization), and SWMM (Stormwater Management Model) are used to assess the impacts of development on stormwater quality and evaluate the effectiveness of different management strategies (Burns et al. 2012, 233, 234; Fletcher, Andrieu, and Hamel 2013,

272). Other models, such as Aquacycle and UrbanCycle, have been developed to represent an integrated urban water cycle that identifies gaps in current strategies. The problem with many of these complex models is the lack of data to help understand the interactions between components of the water cycle (Fletcher, Andrieu, and Hamel 2013, 272–273).

Other emerging considerations are the increase in newly identified micropollutants, the effect of loss of evapotranspiration in urban areas, and the impact of climate change on urban hydrology. Techniques to restore evapotranspiration are being used for a more integrated approach, but improved tools to assess effectiveness are needed. Saltwater flooding is a new concern for water management for low-lying coastal cities because of sea level rise. Some cities are adopting new "waterproofing" protection and accommodation strategies, such as development ordinances that include elevating new buildings and structures, constructing barriers to hold back seawater, and constructing estuary-type wetlands on city perimeters with canals to force seawater around the built areas. These strategies are not "one size fits all," and careful consideration of the options, including the environmental, social, and economic costs, is important. Cities are also adopting new stormwater policies and construction ordinances that consider the social aspects of urban waters. Decisions about urban water policies now reflect the perceived value to humans, and while flood control has an obvious value, social values are now recognized, including the ability to interact with water and nature and enjoy the aesthetics of pleasurable water views. Cities have long recognized that good water views have great value for development, economics, and tourism, which creates incentive to protect and improve the water quality. The science of urban hydrology requires a greater range of disciplines to participate in research on urban hydrology. Citizen scientists can participate in data collection programs to improve the quality of urban water in a variety of ways. Water quality and invasive species monitoring are typical activities for water programs. Most programs partner with government agencies, such as the department of natural resources, and local university researchers, particularly those with extension programs at land grant universities. For example, the Cornell Lab of Ornithology, at a land grant university, coordinates Citizen Science Central, which houses the Volunteer Water Quality Monitoring National Facilitation Project. The project is run by USDA Cooperative State Research, Education, and Extension Service, University of Rhode Island Cooperative Extension, Salish-Kootenai College, and University of Wisconsin Extension.

Citizen Science Program: Wisconsin Water Action Volunteers

Water Action Volunteers (WAV) is a statewide program for citizen groups who are interested in monitoring the health of their hometown streams. Level 1 participants monitor temperature, dissolved oxygen, stream flow, and transparency (turbidity) of the water. They also assess streamside macroinvertebrate habitat and aquatic invasive species. Level 2 participants also measure total phosphorus, chloride, specific conductance, and *E. coli* bacteria. Level 3 projects are determined annually, and volunteers are required to complete Level 1 or some other stream monitoring experience. Data collected by the volunteers are entered into a web-based database, and the program also offers water-related educational materials to educators (WAV 2007).

Case Study: Water-Smart Parks

Water-smart parks use a variety of strategies to move, store, clean, and use stormwater runoff as an amenity and for irrigation. The Trust for Public Land (TPL) supports "close-to-home parks" in or near cities and through its Center for City Park Excellence (CCPE) has published a report about ways parks can help control urban stormwater. The preferred method is the "sponge city" approach to infiltration, but the challenge is finding enough unbuilt land to capture surface water. Parks with pervious surfaces and large spaces are the obvious choice, but most capture only direct rain or snowfall, which means significant infrastructure changes to capture and treat runoff. For park agencies the goals are to improve ecological function of the park hydrology and to save money on irrigation, which can be substantial for some cities: San Diego pays a little over $12 million annually to irrigate 46,168 acres, and Denver, which has 5,957 acres (constituting 8% of the city area, the median for the 100 largest U.S. cities), pays almost $3 million annually for park irrigation. A new generation of stormwater management techniques are being employed, including phytoremediation (using plants to take up pollutants), green parking lots with engineered or structural soil, and constructed facilities such as sports field detention areas, and underground cisterns (The Trust for Public Land 2016).

5

Urban Vegetation

Urban vegetation refers to plant material that is both cultivated by humans and naturally established in the built environment of the city. While not normally thought of as natural habitat, cities with vegetation are considered habitat by some ecologists because they contain nature, host biodiversity, and provide ecosystem services. It is hard to imagine a city without plants for the beauty they provide, but plants also enhance livability for city dwellers with ecosystem services, including improved air and water quality, healthier soils, and moderation of extreme temperatures (fig. 5.1). A renewable resource, plants also sequester carbon, support wildlife, provide food, increase property value, and protect us from severe weather such as hurricanes and floods. Plant communities (a collection of plants) in cities can be very different in structure from plant communities in natural areas. Adaptability and self-selection create landscapes in natural areas, while visual preferences, plant availability, and landscape policies and ordinances create landscapes in the city at both the macro and micro scale. For example, macro scale urban forest management policies typically apply to the entire city, street tree ordinances apply to special districts in the city (fig. 5.2), and HOA (homeowner association) landscape codes apply only to a neighborhood and more specifically, at the micro scale, to each yard in the community.

Livability, that distinct quality that makes residents feel healthy and happy in their communities, is greatly enhanced by vegetation, especially by trees in neighborhoods and parks (fig. 5.3). Experts put a high value on trees for their ecosystem services, but most residents love them for the beauty, shade, and wildlife habitat they provide. The concept of monetary value of trees and other vegetation was explored using a new tool called i-Tree Canopy, to survey the tree canopy in 10 megacities around the world to determine the value of the benefits of urban trees, estimating a median benefit of US$482 million annually, or $35 in free services to each resident

FIGURE 5.1. Coastal village, Turkey: diners enjoy the shade of vines that make outdoor dining comfortable and aesthetically pleasing.

FIGURE 5.2. Paris, France: pedestrians enjoy the shade of street trees on the Champs-Élysées, an iconic boulevard in Paris.

FIGURE 5.3. London, England: a row of pleached trees adds character and creates a pleasant walking path.

(Endreny et al. 2017, 329). Citizen science projects have also been utilized to determine tree value. One example is Cool Green Science, a project that helps track the value of trees using i-Tree Streets and i-Tree Pest Detection, to estimate the tree's effect on air quality, greenhouse gases, and stormwater protection. The information is used to build a national database for managers to track trends in pest sightings (Feldkamp 2015).

Until recently, research on urban ecology and urban vegetation was nonexistent as a field of study, which is often attributed to the reluctance of ecologists to study "unnatural" systems. However, a new research agenda is emerging with the recognition that vegetation in cities is a prominent and highly valued feature. The complexity of the urban environment requires a different approach to the study of urban plants. A socioenvironmental focus is needed with input from ecologists, biologists, entomologists, and horticulture and arboriculture experts. Others, such as landscape architects, engineers, developers, and policy and land management experts are needed to apply the research to policy and development.

What: Characteristics of Urban Vegetation

Urban plants, or flora, can be intentionally planted or spontaneous (growing naturally), and they include native and non-native annuals, perennials,

FIGURE 5.4. Monterey, California: a tree stump tangled in a chain-link fence. Fences can provide initial support for trees but become a problem as trees grow.

herbs, shrubs, and trees. Urban vegetation can be found in many areas, including public and private green spaces, as streetscapes along transportation networks, in schoolyards and recreation areas, in community gardens and residential yards, and in abandoned properties and cracks in sidewalks. Plants in most urban green spaces are selected and planted to create an intentional landscape, one that combines texture, form, and color for aesthetics and functional plants for recreation or other uses.

Other plants are spontaneous, growing naturally where conditions are favorable with water and sunlight. This "other" flora are the stubborn plants, sometimes known as weeds (defined as a valueless plant in the dictionary; "weed" is not a biological term), that appear in cities to create spontaneous little landscapes with no apparent purpose or design (Seiter 2011). Spontaneous vegetation, also known as SUPs (spontaneous urban plants), is often found in specialized urban habitats where structures, hardscape, and microclimates support plants in unique ways. Examples of special habitats include sidewalk cracks, chain-link fences (fig. 5.4), stone walls, street medians, abandoned lots, and river and railway corridors. The amount of spontaneous vegetation in a city often correlates with lack of economic prosperity, because the conditions in economically depressed cities with more abandoned land allow more time for succession and establishment of spontaneous plant communities (Del Tredici 2014). Despite providing ecosystem benefits similar to cultivated plants, weeds are strongly discouraged in urban areas and are often incorrectly labeled as invasive plants. Weeds can be native plants (typically 30%–50% of plants in cities are native), but they are mostly non-native cultivars that grow aggressively, an adaptation that makes them well suited for tough urban conditions. In fact, urban areas often serve as a refuge for establishment of non-natives, which can then serve as a source for further invasion both outside and within the city (Dunn and Heneghan 2011, 104).

The diversity of plant material in urban areas is driven as much by socioeconomics (political and social inequality), as environmental conditions. Measures of social status, income, and wealth are correlated with plant diversity. Wealthier neighborhoods often have a higher number of different tree species, and where homeowners earn above the median annual income, plant diversity is more than double the neighborhoods below median annual income. Similarities in species composition and richness have been noted in cities that are in close proximity. Availability of plants, management of gardens, and socioeconomic status appeared to be the reason, rather than factors related to the local climate or environmental conditions of each city (Dunn and Heneghan 2011, 109, 113).

Challenges to planting and maintaining vegetation are numerous in cities, including lack of healthy soil, high temperatures from the urban heat island effect, lack of water due to sealed surfaces, air pollution, and lack of areas to grow plants. Establishing and maintaining healthy plant material requires an understanding of how these conditions affect plants and methods to mitigate the impacts. Intense land management practices commonly used in cities, including mowing, applying fertilizer, and irrigating also tend to lower plant species richness. The use of plants in urban areas is often controlled by county or city policies, such as urban forest management plans that include street tree and parking lot codes, and landscape maintenance practices by recreation and parks departments for recreation sites. Neighborhoods are also often subject to landscape codes and covenants set forth by homeowner associations.

So What: Spontaneous Vegetation and Management Concerns

Most people agree that the ecosystem services provided by plants in urban areas are beneficial; however, they can also come with a set of unique problems that are not normally found in natural areas. Most of the environmental problems associated with urban plants stem from the type of maintenance required to keep plants healthy in less than ideal situations. Poor soil quality and compaction, constricted areas for growing, construction debris, underground utilities, and unnatural light cycles create conditions that often require human intervention to keep plants healthy. Chemical contamination, a common problem caused by human intervention, occurs primarily when heavily compacted soil cannot absorb water, causing fertilizer runoff when heavy rains or too much irrigation water sheet-flows across the surface into water bodies. Poor-quality soils that lack nutrients for healthy plant growth create landscapes that are more susceptible to pests and competing weed plants. The subsequent treatments with pesticides, fungicides, and herbicides contribute to the chemical mix that degrades soil and urban water bodies.

Recently turfgrass has become the urban plant of concern, with the frequency and timing of fertilizer applications and irrigation schedules becoming policy dilemmas in many cities. A sharp division between "pro- and anti-turfers" has emerged, sometimes fueled by conflicting research on the contribution of fertilizer to the growing problem of algae blooms in urban waters. When used properly, fertilizer can be an important component of good management practices. However, a desire to adhere to the social norms to maintain a green lawn, community landscape codes that require green turf, and pressure to sell maintenance services to homeowners all drive the overuse of fertilizer. Turfgrass receives much of the blame for water quality and quantity problems because it is ubiquitous in urban landscapes and by some estimates covers more acreage than all the corn grown in the United States (Diep 2011). Besides the fertilizer issue, turfgrass also requires more water than drought-tolerant landscape plants. However, the argument about turf can go both ways; healthy turf can prevent erosion, cool surface temperatures, sequester carbon, and keep soil healthy by improving water and air penetration with deep root growth. Turf also has an important social and functional role as a tough, trample-resistant ground cover in parks, residential yards, and sporting venues to support games and recreational activities. Landscape designers and homeowners also appreciate turf for its aesthetic appeal; many people like the look of an open, well-maintained lawn that provides contrast to the sometimes-chaotic look of shrubs and bedding plants and creates a visually pleasing composition. Behaviorists also hypothesize that the appeal of turf can be attributed to a hard-wired evolutionary desire for a simple

FIGURE 5.5. Monterey, California: a tiny crack between the building and sidewalk provided a place for a shrub to take root.

landscape that is easy to "read" and understand, which enabled a better chance of survival for our ancestors (Kaplan 1987).

Weeds that appear in turf, sidewalks, buildings (fig. 5.5), and driveways are also a contentious issue in urban landscapes. An expectation to keep lawns and outdoor areas weed-free, especially in neighborhoods with landscape covenants, promote the overuse of herbicides. In many home-owner associations, the most common landscape condition that prompts warning letters to homeowners is the admonishment to keep their lawns weed-free, so as not to decrease property values and spread weeds to their neighbors. In the past, weeds were a concern mainly in agriculture, where they compete for nutrients, water, and light and reduce crop yield. In the urban environment they compete for the same resources, but rather than crop yield, the concern is aesthetic quality of the landscape and preventing their spread. Weeds are plentiful because they have characteristics that allow them to survive in harsh conditions, such as abundant seed production, seed dormancy with long-term survival of buried seeds, and quick establishment with rapid spread. However, in spite of some negative impacts, weeds do have benefits, including soil stabilization, stormwater mitigation, increasing organic matter in soils, and providing food for wildlife and nectar for bees (Ligenfelter 2009). A major concern for the public with weed control is the effect of the chemicals used to control them on water quality, public health, and nontarget species when the herbicide is allowed to drift or run off site (Wilen 2018). Weed control in plant beds is difficult because a variety of plants make selecting herbicides challenging, resulting in an overreliance on the use of glyphosate and other nonselective post-emergence herbicides (Marble and Koeser 2018). Because of its pervasive use, glyphosate has recently been implicated in legal actions as a cancer-causing substance for landscape contractors and other people who have frequent contact with the chemical in routine work applications.

Although a perceived lack of biodiversity exists in cities, studies show otherwise. Certain areas have remarkable diversity, often attributed to human preferences and activities, including aesthetic values that show preference for a variety of plants and the large number of plant species available to consumers. However, biodiversity can also be lowered by overuse of certain plants by landscape companies, overstocking or limited availability of plants in nurseries, restrictive landscape covenants, and poor growing conditions. New and innovative programs prompted by a growing conservation ethic and changing standards of visual quality are beginning to address some of these problems.

Now What: New Technology and Philosophies for Urban Vegetation

A new philosophy about weeds in urban areas is being promoted to gain more acceptance and appreciation for spontaneous plants that thrive in urban places. In a campaign to bring attention to weeds on sidewalks and tree pits in Brooklyn, New York, a team from Future Green Studio circled plants with bright yellow paint and identified them with a placard. The goal was to have people realize that plants that can maintain themselves in extreme conditions have ecological benefits, and rather than eradicating them we should look for opportunities to encourage their use. To inspire a different perspective, the team catalogued the weeds in "Profiles of Spontaneous Urban Plants" with specimen-quality photos in the tradition of botanical illustrations to change the context and elevate the status of the plants (Seiter 2011). In summer 2014 they launched a user-generated database of weeds in cities. The intent is to encourage debate among the general public, designers, ecologists, and artists about the stigmas and societal perceptions of weeds. The documentation has generated discussion about the relevance of labeling plants as native, invasive, or non-native in urban areas (Seiter 2015).

Limited space on the ground to grow plants has prompted new and innovative ways to grow plants in and on structures. Sometimes called living architecture, green walls, green roofs, and living spheres are not new, but new technology is creating other uses for these green spaces (see chapters 13 and 18 for further discussion). A new technology that harnesses the breakdown of organic matter used by plants and converts it into electricity is being used to turn green roofs into power plants to charge phones, power lights, and reduce reliance on other electricity sources. The technology, developed at Wageningen University, converts waste electrons and protons generated by soil bacteria into usable electricity. The company, called Plant-e, also offers products to charge mobile phones in parks and lights in street roundabouts (Ingham 2014).

The Salk Institute is taking a new approach to solving environmental problems associated with climate change with their Harnessing Plants Initiative (HPI). The HPI goal is to discover ways to optimize a plant's ability to capture and store carbon and adapt to different climate conditions. The presumption is that, by helping plants store more carbon, we can mitigate climate change effects and at the same time provide more food, fiber, and fuel to people. By understanding a few genetic pathways in plants, the plant biologists at Salk believe they can help plants grow a bigger root system to absorb more carbon and bury it in the ground in the form of suberin (cork), a naturally occurring carbon-rich substance found in roots. They are using genetic and genomic techniques to develop Salk Ideal Plantsä. The plan is to transfer the genetic traits to crops such as corn, and to coastal and aquatic plants, to protect plants from the stresses of climate change, and to add more carbon to the soil (Salk Institute 2019).

In a different take on the concept of networked trees (trees sending chemical, hormonal, and electrical impulses through underground fungal networks to each other), officials in Melbourne, Australia mapped all 77,000 trees in the city and assigned each an ID number and email address as part of their urban forest strategy. The purpose is to encourage people to report problems to facilitate management of the trees. The quirky unintended consequence was that people wrote thousands of letters to the trees, thanking them for all they do for people, professing their love for them, and sometimes just to say hello to the tree. While the email program revealed the love of city dwellers for their trees, the program also encourages civic engagement and helped with maintenance of the trees; as a result, 12,000 new trees have been planted since the program started (Ley 2015). The biggest challenge in assigning ID numbers and email addresses is knowing where trees are located. Mapping city trees is an enormous task, sometimes taking years, but new technology using artificial intelligence is making that task much easier. Cartographers and scientists have developed a machine learning model that does not have the limitations of current methods to differentiate between different types of plants. Currently, NDVI (Normalized Difference Vegetation Index) which is derived from satellite imagery, is combined with LIDAR data to differentiate vegetation by height. LIDAR measures height by shooting a light from a drone and measuring the length of light bounce-back. However, LIDAR is costly; to solve the problem of cost, researchers used existing data sets from NDVI and LIDAR to create algorithms that teach the machine learning model to differentiate trees from shrubs and grasses (Poon 2018).

Citizen Science Program: New York City EcoFlora Project

The EcoFlora Project engages the public in protecting and preserving the city's native plant species. After training by experts, the citizen scientists are observing, collecting, and recording data on the plants and their relationships with other organisms, such as insects and birds, and combining these data with known information from scientific publications and collections. EcoFlora is a real-time, online, ongoing checklist of plants intended to be a dynamic resource for conservation planning (NYBG 2019).

Expert Insight: Plant Blindness

Urban vegetation can be exciting and interesting. However, some people, for a variety of reasons, have no interest in plants and simply ignore them, a behavior labeled "plant blindness." While ignoring plants may seem like a harmless behavior, preventing plant blindness could be a key endeavor for protecting and enhancing our urban landscapes. Plant blindness is defined by several behaviors, including inability to recognize plants, lack of appreciation for their beauty and unique biological features, and inability to recognize the importance of plants to humans, judging them as inferior to humans and animals and not worthy of consideration. Persons experiencing plant blindness don't pay attention to plants in their daily life and don't consider the needs of plants. Many lack hands-on experience with plants and knowledge about plant life. Characteristics of human perceptual, intellectual, and visual cognition may explain why some people overlook plants. Perceptually, plants lack meaning to the observer, which may be due to lack of education, so they are less likely to notice plants. Intellectually, plants are generally nonthreatening, so most people can ignore them with no concern for consequences. Visually, our brain works to make sense of our surroundings by detecting differences, so when masses of plants blend together the lack of contrast means they are simply not noticed. One suggestion to overcome plant blindness and generate more interest in urban plants is to recognize the tendency of instructors to put more emphasis on animals (zoocentrism) in introductory biology courses and encourage classes to have more plant-based content (Wandersee and Schussler 1999, 84–86).

6

Urban Wildlife

Urban areas support an amazing variety of animals, including wild and domestic, native and non-native, and invertebrates and vertebrates. Nearly every taxonomic group of insects, amphibians, reptiles, fish, birds, and mammals can be found in cities. While most urban animals live in green spaces including yards, parks, and golf courses, some occupy buildings, others live under bridges and highway overpasses, and some, such as the sacred cows in India, live in the streets. Other tiny animals, such as head lice and fleas, live on humans and other animals (Douglas and James 2015f, 240). Some animals are welcomed by people, others are considered a threat, and others go completely unnoticed, adept at hiding, or so tiny they can't be seen. Working animals, such as police horses and guide dogs, are common in cities (Douglas and James 2015f, 239), and now a relatively new category of working/domestic urban animals are found on urban farms, including chickens and goats.

Wildlife science, a field of study concerned with animals in natural habitats, has only recently begun to consider urban areas as natural habitats for some animals. Initial research in urban areas was concerned primarily with hazardous wildlife encounters in cities, such as bird fatalities in window collisions and roadkill counts, but new studies are focused on managing wildlife and studying the impact of the built environment on wild animals. The study of urban wildlife is multidisciplinary and includes experts in anthrozoology, who study the interactions between humans and animals, including animal ethics, welfare, and law. Other experts include wildlife biologists, zoologists, and specialists in entomology (insects), herpetology (reptiles and amphibians), ornithology (birds), mammalogy (mammals), and ichthyology (fish). Habitat experts include those who study the ecology of urban areas as it relates to wildlife, including urban forests, wetlands, and parks. Other fields include wildlife conservation, veterinary science, evolutionary biology, and animal behavior experts.

Citizen science programs for wildlife have historically focused on bird counts and location sightings, some in urban areas, but most in natural areas. A new citizen science program, City Nature Challenge, was one of the first to focus on urban areas. The BioBlitz project documented nature in more than 69 cities around the world from April 27 to April 30, 2018, and again in 2019 and 2020. The challenge is an international effort to document wildlife and plants in cities around the globe, with a goal to create a database and promote biodiversity awareness and engagement. Participants upload photos tagged with the photo's location on the iNaturalist app. In 2018 there were 428,401 observations worldwide of 18,418 species by 17,380 observers. The challenge is organized by the Natural History Museum of Los Angeles County and the California Academy of Sciences (Ueda 2018).

What: Characteristics of Urban Wildlife

Generally urban animals have been grouped into three categories; the first is urban exploiters: animals that can adapt but are almost completely dependent on some human help. They usually live in and around buildings and are often considered pests, including mice, rats, and some birds. Second are urban adapters; these are native species that thrive in urban areas and can exploit resources, including many birds (fig. 6.1), and squirrels, raccoons, opossums, and coyotes. The third category is urban avoiders—animals that cannot adapt and typically leave as humans move in, including mountain lions, bears, bison, and elk (Douglas and James 2015f, 238–239, 245; Jarvis 2011, 353). Some characteristics that allow the urban exploiters and adapters to survive include being generalists (rather than specialists) in food, water, and shelter options, and possessing the ability to adjust to human activity and disturbance. High reproduction rates, having few competitors and/or predators, and having the ability to negotiate fragmented landscapes also help them survive (Adams and Lindsey 2011, 121; Douglas and James 2015f, 243; Jarvis 2011, 358). The adapters will also often adopt different behaviors in urban areas that increase their survivability. For example, compared with their rural counterparts they often reduce their migratory behavior, wait longer to take flight or run when people approach, are more aggressive, tend to be more exploratory, and reproduce earlier (Douglas and James 2015f, 241). The greatest variety and number of species live in parks and patches of native vegetation rather than recently developed landscapes (Jarvis 2011, 352). Cemeteries, private

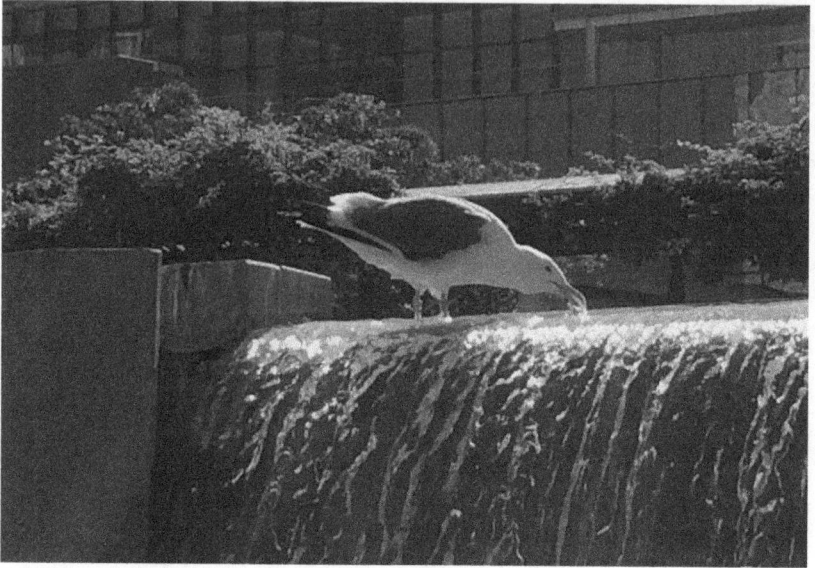

FIGURE 6.1. San Francisco, California: a gull drinks water from a fountain in Yerba Buena Gardens, a large public park.

yards, golf courses, stormwater ponds, rail lines, road rights-of-way, and abandoned lots also support high numbers of animals. Many of these landscapes are found in suburban areas where species abundance and diversity both peak (Jarvis 2011, 353).

While domestic animals in cities depend entirely on humans for food, there are numerous wild animals who supplement their diet with food provided by humans, either intentionally, in bird feeders and lunchtime feedings, or unintentionally in trash bins, pet food bowls, and outdoor cafes. The most successful animals are opportunistic feeders that have figured out innovative ways to get food (Jarvis 2011, 354). For example, the Cooper's hawk has learned to stake out bird feeders in Chicago for easy prey. A research team used data from the Citizen Science Project of Cornell Lab of Ornithology to determine the extent of predators at feeders. The participants recorded birds that showed up at feeders and noted how long predators were watching and then catching birds (Kobilinsky 2018). The ability to find food is the most important predictor of survival in urban areas. A historic example is the house sparrow, a very common bird in late Victorian Britain where horse-drawn carriages were popular. The sparrows fed on abundant supplies of horse grain, but when motor vehicles became popular, the sparrow population dropped dramatically

(Jarvis 2011, 356). While food is important for survival, it is usually not a limiting factor in life expectancies. Urban animals typically have shorter life spans because of more accidental deaths, including death by domestic cats or dogs, electrocution, collisions with windows and buildings, drowning in swimming pools, poisoning, and intentional killing with traps or shooting (Douglas and James 2015f, 241; Jarvis 2011, 356). Certain animals that are vectors of diseases harmful to humans, such as insects (mostly mosquitoes) and rodents, are often the target of citywide campaigns to eradicate them.

Public attitudes about urban animals vary with the species; some are enjoyed and supported, such as squirrels, ducks, birds, and butterflies, while others, including rats, mice, and cockroaches, are often exterminated. The ten vertebrates of greatest concern in American cities, based on greatest to least magnitude of damage or danger, include the raccoon, coyotes, skunks, beavers, deer, geese, squirrels, opossums, foxes, and red-winged blackbirds. The raccoon ranks highest because their behavior as scavengers leads to complaints of noise, damage, and the possibility of being vectors of rabies (Jarvis 2011, 356).

So What: Conservation and Social Issues with Urban Wildlife

Cities have both negative and positive impacts on wildlife. Most ecologists would agree urban habitat impacts on wildlife are mostly negative; examples include little food value in exotic plant species, fragmented habitat and road migration barriers, and lack of ecosystem stability with soil disturbance and chemical use. The result is change in competitive relationships among animals that favor those that adapt easily (Adams and Lindsey 2011, 116–117). Different species respond to urbanization in various ways. For example, nighttime light pollution will disturb some bat species, while amphibians are impacted by chemicals in urban waters and loss of aquatic habitat. Carnivores can be poisoned by eating rodents that have accumulated rodenticides, and snakes often suffer severe cuts and entrapment when tangled in plastic landscape netting. Some species are at risk of disease by interacting with domestic pets, and others may have trouble finding mates because of decreased populations (The Urban Wildlife Working Group 2012).

Insects and mammals top the list of conflicts between humans and wildlife, including problems with mosquitoes and cockroaches, and with raccoons, coyotes, skunks, beaver, and deer. Many of these same species are problems internationally, but that list also includes monkeys, pigeons, wild boar, iguanas, and gulls (Adams and Lindsey 2011, 120). Of particular concern for public health are zoonotic diseases, parasites, and viruses that are often easily passed to humans. For this reason, management activities usually include population control and modifications to animal and human behavior. This sometimes-unrealistic goal is largely dependent on changing human behaviors, such as supplemental feedings, that often drive up population numbers (Adams and Lindsey 2011, 121; McCance et al. 2017, 8). Hazardous wildlife include those that are disease vectors or poisonous, such as mosquitoes and snakes, those that can destroy buildings, such as termites, and those that are simply a nuisance, including deer and pigeons. Insects are the most common animal in cities, and of all the species of urban wildlife, they cause more damage, transmit more diseases, and are the hardest to control; yet they are also the most necessary. As cornerstones of the living world, they pollinate plants, help disperse seeds, build healthy soil, and help decompose dead plants and animals. Insects also help control weeds and other insects, provide high-protein food, and are used to create drugs and medicines. They also contribute about US$57 billion annually to the U.S. economy (not including pollination) as the base of the food chain for wildlife and humans (Hoff 2018).

Managing wildlife in urban areas poses more challenges than natural areas. Urban management is often called a "wicked" problem because it takes place in a social-ecological system that involves a complex set of sociocultural, political, economic, and ecological components (McCance et al. 2017, 2). Few state or federal programs provide funding for urban wildlife control, and the layers of jurisdiction in urban areas pose a problem. Fewer wildlife ecologists are trained to manage urban animals, and the attitude of the residents present challenges; some love wildlife, others do not. Often dealing with conflict among the people involved is more challenging, including the expectations people have of benefits they will experience; for example, the ability to view or interact with wildlife in city parks or new interactive zoos (fig. 6.2). While people often expect management to create opportunities for physical and social enjoyment of wildlife, management's goal is to make sure the interaction is positive, so people are more accepting of wildlife. Human values, based on strong emotions toward wildlife, are major drivers of management practices that govern the interaction between people, wildlife, and habitats (McCance et al. 2017, 1–2, 6–7). For many residents, the city is where they have their first experience with wildlife, and the opportunity to observe animals can improve

FIGURE 6.2. Perth, Australia: Kangaroos in an exhibit at the Perth Zoo. Kangaroos roam freely in a fenced area for visitors who walk among the kangaroos.

social interaction, such as bonding over shared experiences. Animals can also improve health and reduce stress, usually through the act of observing them, which also motivates people to care more about nature. Concern for nature often prompts people to change their environmental behavior, such as using fewer pesticides and chemicals and using more native plants in their gardens.

Now What: Managing Wildlife in Urban Areas

Strategies to improve urban habitat for wildlife vary depending on the species. Approaches include reducing bird strikes with green buildings, building more green roofs, installing nest sites on buildings and utility poles, and establishing native plant gardens for pollinators that include backyard ponds and bat and bird roosts. Green infrastructure strategies include maintaining important natural areas and using green streets and sustainable stormwater management for aquatic habitats. Large scale, citywide

FIGURE 6.3. San Francisco, California: Harbor seals on a platform in the bay. Designed as a resting spot, the platforms also offer visitors a close view of the seals from the piers.

strategies include establishing a strong urban tree canopy, protecting riparian corridors and floodplains, creating wildlife corridors with safe road crossings, and providing education and outreach to citizens. In one citizen science project, volunteers operated more than a thousand cameras in green spaces near their homes and found that cemeteries and golf courses in suburban areas have more species variety than parks with mature trees and turfgrass, making them good models for green space design (Gallo and Fidino 2018). To help urban planners and landscape designers incorporate the needs of wildlife into their design projects, the 2018 Greenways for Wildlife Project used design terminology to develop model design guidelines, including parameters such as corridor width and adjacent development intensity. (The Urban Wildlife Working Group 2012). One concept used by designers is "shared spaces" for human-wildlife coexistence that enhance viewing of wildlife (fig. 6.3). The key is to consider the social-ecological carrying capacity, which is the ability of the space to meet the needs of each and support both. Conditions for a shared space should include defining acceptable human interactions with wildlife, developing norms or standards for the space, and ensuring a favorable (mostly positive) balance of negative and positive human-wildlife interactions. The goal is to improve risk perceptions and reduce negative experiences with urban wildlife so that individuals and communities see value in habitat conservation and urban wildlife (McCance et al. 2017, 13–14).

Professional groups such as the National Wildlife Control Operators Association (NWCOA) are advancing more ecologically sound methods for controlling nuisance wildlife. Several states now require a license for wildlife control operators (WCOs) to remove animals, and the National Wildlife Control Training Program is a basic course for licensure. WCOs try to humanely remove animals, use toxicants only when necessary, or modify animals' behavior with repellents or barriers. The association is developing methods to encourage cooperation and coordination between jurisdictional agencies to address boundaries that are ecologically connected even if they are not politically or programmatically connected. Activities include developing strategies for urban growth planning and cost-share funding for conservation planning. One objective of conservation planning is to develop innovative restoration techniques such as managing stormwater to mimic natural hydrology, using native plants, and restoring historically important habitats. Another goal is to build partnerships between fields in urban ecology and environmental social sciences for research and monitoring, including studies in habitat restoration and land planning with education and outreach (The Oregon Conservation Strategy n.d.). Outreach activities can include citizen involvement in restoration projects, creating more opportunities to view wildlife, giving instructions on clearing brush and using pesticides, and suggestions to mitigate the impacts of cats, both domestic and feral.

New approaches to protect and increase insect populations include providing habitat corridors, such as the I-35 Monarch Highway from Texas to Minnesota, where the roadsides are planted for monarch butterfly habitat, managing urban parks and forests in more ecologically friendly ways, and educating people about the ecological benefits of insects as service providers rather than pests. An example is research in which scientists are pairing venom from deathstalker scorpions with fluorescents to help see brain tumors during surgery (Worrall 2017). Huge potential exists for citizen science to contribute to assessing the status of insects. Once nonscientists learn to identify insects, they can generate large amounts of data. Several citizen science projects with insects already exist, including Dragonfly Migration Monitoring, Vanishing Fireflies Project, School of Ants, National Cockroach Project, and ZomBee Watch. Activities include tracking bumblebees and dragonflies in North America, counting overwintering monarch butterflies in California, and counting fireflies on the U.S. east coast (Hoff 2018). Citizen science programs are also active in monitoring many types of wildlife; some programs include Frog Watch, Wildlife Tracker, Road Kill Survey, Celebrate Urban Birds, Bat Detective, and The Wildlife of Your Home Project.

Citizen Science Program: iNaturalist Mobile App

The iNaturalist mobile app is a citizen science project developed in 2017 with the California Academy of Sciences and the National Geographic Society. The worldwide online network has millions of citizen science observations of plant and animal biodiversity. The goal is to crowd-source species identification and record organism occurrence to understand when and where organisms occur. Photos and location are uploaded, and the information is shared with scientific data repositories like the Global Biodiversity Information Facility to help scientists find and use the data (iNaturalist 2017). One study from the University of Illinois used the app to compare telemetry data from radio-collared coyotes with citizen science data from iNaturalist to track negative interactions between people and coyotes. Although there was a difference between the data—the collars showed coyotes using a large arboretum vs. the app showing them in developed areas—the data was used to help manage potential conflicts (Kobilinsky 2019).

Expert Insight: Feral Cats and TNR Programs

Feral cats are a good example of an animal population that has adapted to urban conditions with the help of humans and as a result has become a serious management and social problem with advocates on both sides of the issue. Feral cats cause many economic and health problems for humans and ecological disruption for other wildlife. Although exact numbers are unknown, some experts estimate hundreds of millions of birds and more than a billion small mammals are killed annually by domestic and feral cats (Mott 2004). Other experts note that birds in urban areas are also killed by habitat loss, pollution, window strikes, and pesticides. Feral cats carry diseases such as toxoplasmosis (a parasitic disease), rabies, and cat scratch fever, all problematic for humans and wildlife. They have been listed among the world's worst urban invasive species globally (American Bird Conservancy n.d.). Efforts to control feral cats include Trap, Neuter, and Return programs (TNR), but their efficacy is questionable; while they may reduce numbers, they don't solve the problem of bird and small mammal killings, because even well-fed neutered cats continue to hunt. However, advocates of the program believe it is the best way to gradually and humanely reduce the population. Although some people advocate holding those who man-age cat colonies liable for the loss of birds (especially protected species), a feral cat would have to be classified, in biological and legal terms, as a domestic animal (chattels of their owners) as opposed to a "wild" animal, to enforce liability. Most communities do not have ordinances holding cat owners liable for injuries (most are specific to dogs). However, conserva-tion laws, such as the Endangered Species Act and the Migratory Bird Treaty, apply without regard to ownership and could make claims action-able. However, most legal minds agree that passing ordinances that control how feral cats are fed would yield more significant results (LaCroix 2006).

7

Urban Climate

Urban areas are well known for their towering buildings, variety of architectural materials, and busy streets. All of which add to one more thing cities are known for—hot summer days. High temperatures and humidity, caused by the urban phenomenon known as urban heat islands (UHIs), are recognized as a significant environmental condition that impacts livability in communities. The difference in temperature between urban areas and the surrounding rural areas is what defines the urban heat island. In a city with a million or more people the annual mean daytime air temperature can be 2–5°F warmer than the surrounding rural area, and the nighttime difference can be as much as 12–22°F warmer (USEPA 2008a, 1). The impact of UHIs was highlighted recently when as many as 30,000 people died in heat-related deaths in the 2003 heat wave in Europe, where many buildings lack air conditioning and cities have no policies for mitigating heat stress emergencies. It is estimated that if greenhouse gas emissions continue, cities will experience twice the number of days that feel hotter than 100 degrees by 2050. Air pollution is also a factor in urban heat islands; particulates emitted in the heat dome help trap more heat and block the release of heat to the upper atmosphere, intensifying the health hazards of the UHI. The World Health Organization estimates that globally, 1 in 4 deaths are related to air pollution, including 4.2 million deaths annually from outdoor air pollution (WHO n.d.). In addition to heat-related health problems, increased high-heat days have a direct impact on the economy of many cities as they struggle with energy shortages, business disruptions, declines in tourism, and emergency health crises.

Urban climatologists study climate in cities, but the complexity of urban heat islands requires input from other specialists, including meteorologists, soil scientists, ecologists, and electrical, civil, and environmental engineers. Construction and building materials experts work with urban planners and landscape architects to create heat-resistant cities, and social and public health researchers are needed to develop crises strategies. Planners and developers are looking for evidence-based research from many fields to help develop mitigation and adaptation strategies. Issues they are dealing with include the use and effectiveness of new "cool" building materials, new redevelopment and urban form strategies, the efficiency of green roofs and green streets, alternative power sources to protect against power outages, strategies for heat emergencies, and the use of heat mapping data to develop prediction models and intervention policies. Citizen science projects usually center around urban heat island mapping campaigns and collection of air samples to assess air pollution locations. Heat mapping projects have been organized by the National Oceanic and Atmospheric Administration (NOAA), the Climate Program Office (CPO), and the National Integrated Heat Health Information System (NIHHIS) to help develop standards to build heat resilience in cities. Heat maps locate temperatures in thousands of locations in the cities to help determine the causes of extreme heat areas and to look at other data associated with these points, such as health records, population demographics, air quality, landscape features, and number of first responder visits for heat-related emergencies. The goal is to use the information to help reduce negative health impacts and develop planning policies to mitigate the heat island impacts (NOAA 2018).

What: Characteristics of Urban Climates

Urban climate refers to the climatic conditions of urban heat islands caused by the built form and the economic, industrial, and social activities of the city. When cities are built, they replace vegetation and open land with roads, buildings, and other impermeable hard surfaces, creating a hot, dry environment unique to cities. The microclimate conditions modified by city form include temperature, wind, rain, and humidity (fig. 7.1). UHIs are caused by several factors, including the overall shape and form of the building mass, building materials that store and conduct heat, heat from traffic and utilities, loss of vegetation and increased impervious surfaces such as roads, increased particulate pollution in the air, and the number of open areas and water bodies within the city (Douglas and James 2015b, 80–85). Building and road materials commonly used in cities such as concrete, stone, steel, brick, and asphalt have high heat capacities, making cities efficient at storing heat in their infrastructure. In addition to buildings as heat sinks, the surface texture of buildings also affects the velocity and

FIGURE 7.1. Paris, France: typical street where tall buildings create a narrow "street canyon." The buildings, trees and road surface modify climate in cities.

direction of wind flow, causing turbulence in narrow streets flanked by tall buildings (also called urban canyons), adding to the UHI effect (Douglas and James 2015b, 78–79; Grimmond 2011, 106; USEPA 2008a, 9).

Urban heat islands are characterized by a dome of warm air over a city. The dome has two main layers, the urban canopy layer (UCL) and urban boundary layer (UBL). The urban canopy layer contains all the buildings and roads and rises from the ground to the rooftops. This is the layer where most of the dissipated heat affects the city. The top layer, the urban boundary layer, is the airflow from outside the city (usually from rural areas) that flows over the UCL, carrying heat and pollution downwind from the city (WHO n.d. 3; Douglas and James 2015b, 77, 81).

Temperatures rise and fall in the dome from day to night based on energy flows measured as urban heat budgets. Heat budgets are the balance of incoming and outgoing energy flows, called fluxes, which are characterized by the transfer of thermal energy (heat) from one surface to another or to the atmosphere (Douglas and James 2015b, 81–82; USEPA 2008a, 11). The dynamics of the urban heat budget explain the urban heat island: basically that the transfer of heated air to and from building surfaces does not allow for significant temperature decrease at night to cool the buildings or air (Parlow 2011, 34; Grimmond 2011, 109, 111). Several heat fluxes occur in the urban budget, including sensible, latent, storage, and anthropogenic fluxes. The sensible heat flux is heat sensed by the human body or a thermometer and occurs when radiation from daytime sunlight heats surfaces, is absorbed into structures, moves back to the surface at night, and is released into the atmosphere, creating the heat you feel radiating off a building on a cool night (Grimmond 2011, 113). The latent heat flux, sometimes called evaporative heat flux, is the transfer of heat to the atmosphere through evapotranspiration, usually from precipitation (rain or irrigation), and plant material. Latent heat fluxes are typically not large in urban areas because cities have less moisture as a result of limited vegetation, warmer air temperatures, and stormwater systems that rapidly drain water from surfaces (Grimmond 2011, 112; USEPA 2008a, 7). Storage heat flux is the energy retained in soil or buildings that can change with seasons, and anthropogenic heat flux is the small amount of energy produced by human activities, including heat from the combustion of fuels in cars or manufacturing. A positive heat flux happens when a surface, such as a building or road, absorbs heat, cooling the adjacent air and soil. Negative heat fluxes happen when heat is released from a surface, heating the adjacent air and soil. The result is constant changes in temperature due to positive and negative heat fluxes, especially from day to night, when buildings store heat during the day and discharge heat at night. This day/night difference shows that the urban heat island is due more to a small amount of cooling at night rather than to large heat gains during the day (Parlow 2011, 34; Grimmond 2011, 109, 111).

Building density and height, especially in urban canyons, are the most important factors in nighttime cooling (Oke 2011, 127–128). Early in the day the tall buildings that create urban canyons provide shade, reducing surface and air temperatures on streets; however, when the sun reaches more building surface, they rapidly store solar energy (heat) through incoming radiation. At night the outgoing heat released from the buildings is often reabsorbed by other building or road surfaces, but it never escapes the canyons, preventing nighttime cooling (USEPA 2008a, 10).

Built structures and the heat island also affect wind movement and velocity. As warm air rises above the city center (usually at night when temperature differences are greater), the air above the city moves outward, sinks to ground level as it cools, and flows back into the void created by the rising air, maintaining the wind flow through recirculation of the warm air (Douglas and James 2015b, 90). The buildings, often referred to as bluff bodies, force air flow to move vertically or around the buildings, both increasing wind speed and turbulence. The entire city acts as a regional bluff deflecting air around it, but at the neighborhood or street scale, wind moves in many directions around buildings (Grimmond 2011, 115). Humidity and rainfall are also altered by the built environment, with many cities having more rainfall and thunderstorms than surrounding rural areas. The phenomenon is explained by the extra heat convection that creates cloud development in urban areas. Several factors contribute to convection, including the microscopic particles in air (hygroscopic nuclei) around which water vapor condenses to form droplets. An increase in pollution increases number of nuclei for condensation and the frequency of sensible heat fluxes that often increase precipitation (Douglas and James 2015b, 89). It is important to remember that cities can vary widely in urban heat island patterns, as individual cities vary in architectural form and density, in building materials and economic activity, and in differences in topography and regional climate.

So What: Health Impacts of Urban Climates

The major health impact from urban heat islands is heat stress during the summer. All the atmospheric conditions in urban areas, such as air temperature and humidity, wind speed, and solar radiation, influence the human heat budget. In warm air humans experience latent heat flux and release heat energy by transpiration, from sweating and respiration. When humidity increases, the latent heat flux is less effective, and people suffer more from heat stress. Humans can deviate only slightly from their optimal body temperature (98.6°F) before they begin to experience thermal stress (Parlow 2011, 40–42). Cities that historically haven't experienced extreme summer temperatures are often not equipped with air conditioners to help prevent heat-related deaths, and other cities may not be able to use cooling devices when weak power grids are impacted by high demand. The electricity demand, which increases 1.5%–2% for every 1°F increase in temperature, can overload systems and require controlled rolling blackouts to avoid total power outages. The Centers for Disease Control (CDC) estimates more than 8,000 premature deaths in the United States from 1979 to 1999 due to excessive heat exposure, and it is anticipated that the intensity of summer heat waves will increase with global warming (USEPA 2008a, 15).

Extreme heat also causes thermal water pollution that threatens aquatic animal species. When surfaces such as roads exceed temperatures that are 50–90°F higher than air temperature, the excess heat transfers to stormwater runoff, which raises water temperature in streams, ponds, and lakes. Many species experience thermal shock when water temperature changes by more than 2–4°F in a 24-hour period (USEPA 2008a, 16). The surface layer of soils in cities is often warmer because of the surface energy balance in cities that creates a warm zone under the city. Sealing soil with hard surfaces (waterproofing) also leads to dry soils, called soil desiccation. Dry, warm soils can affect root growth, soil organisms, and inorganic chemical processes. Warmer temperatures also increase decomposition and mineralization of nitrogen, which may affect the sequestration and release of carbon from soil (Oke 2011, 125).

Urban heat islands can have some positive effects in cities with cold climates: the increased warmth improves human comfort outdoors, produces an energy savings in heating of buildings, and limits ice buildup on roads. Extra warmth in the winter may also lead to more rainfall, rather than snowfall, in the winter and more local thermal breeze circulation, similar to a sea breeze in the summer (Oke 2011, 129). Warmth can also create longer growing periods and earlier germination, and warmer nighttime temperatures minimize frost risk, allowing less-hardy plants to survive. A greater mix of vegetation can attract overwintering of migratory birds and favor some insects and wildlife (Oke 2011, 128; Parlow 2011, 38). Microclimates of cities can both reduce and increase the types and amounts of vegetation in built areas, often supporting plants that don't normally grow at the latitude of the city (Douglas and James 2015b, 78–79).

Now What: Strategies to Mitigate Effects of Urban Heat Islands

Various technologies used to learn more about heat island effects include heat mapping and pollution mapping to target neighborhoods for resiliency programs. Heat mapping in cities such as Washington, DC, and Baltimore have been used to concentrate resources for interventions at specific locations (see Citizen Science Program below). The detailed maps can be used to describe certain areas of high heat, especially where vulnerable populations live, to prioritize city planning policies for heat mitigation and public health policies for extreme heat events. Mapping features such as roads, buildings, and trees can help develop models for future city development and redevelopment strategies, such as repaving projects. Forward-looking infrared (FLIR) cameras are used to distinguish cool and warm areas on surfaces such as buildings and green spaces to determine optimal percent of different land covers. Mapping building heights in different areas will also help determine the optimal arrangement and percent of different heights to increase airflow in hot areas (NOAA 2018). The city of Richmond, Virginia used their heat mapping project to develop "Steps to Resilience," a program to provide cooling stations and inform planning decisions (see Case Study).

Strategies to mitigate building warming include using "cool" construction materials with less heat storage capacity, using light-colored road and roofing materials (fig. 7.2) to reflect more incoming solar radiation, and increasing more urban green spaces, especially tree canopies, for shade and removal of pollution particulates. Infrastructure changes include changing the spatial arrangement of buildings and green spaces, using solar and battery "microgrids" to keep cooling centers running if the main grid goes down, and using geothermal power to cool buildings (Peters 2019).

Paved areas cover nearly 30% to 45% of land in urban areas and transfer excess heat to the air, so modifying them could substantially mitigate the heat island. Cool pavements use emerging technologies to reflect more solar energy, enhance water evaporation, and remain cooler than conventional pavement. Cool pavements include those with high reflective quality (light colored), permeable pavements, and grid pavements (metal grids with loose aggregate). Chicago's Green Alley Initiative replaced almost 2,000 miles of alleyways with grid pavement in 2007. Experiments in Tokyo and Osaka, Japan are testing the effectiveness of water retentive porous pavements with a sublayer of materials that hold water that evaporates through capillary action to cool the pavement. Others are testing

FIGURE 7.2. San Francisco, California: white walls and green roof of the new transit building reflect the incoming solar radiation to moderate the building temperature.

water sprinkling—spraying porous pavement occasionally during the day to cool it by evaporation. Both methods have shown good results. Micro-surfacing with a thin polymer resin with reflective sand and other fillers has also been used to increase reflective quality of pavement (USEPA 2008d, 2, 8–9, 13). Cool roofs use the same principles of reflectance and evaporation to reduce heat. Cool roof coatings resemble a thick paint and have additives to improve durability, suppress algae and fungi, and shed dirt (self-wash) under normal rainfall. Other cool roof materials include single-ply membranes, color tiles, and painted metal roofing with "cool colors" that contain pigments that reflect from 25% to 70% of infrared solar energy (USEPA 2008c, 7). New York City has coated more than 10 million square feet of rooftops with a white reflective coating to keep buildings cooler, and Los Angles has developed the concept of cool streets, combining cool roofs, cool pavement, and tree planting to reduce temperatures by as much as 20°F compared with conventional streets (Peters 2019).

Another strategy is to cover cities with trees and buildings with green walls and roofs (fig. 7.3). Cities around the globe, including Melbourne, Milan, Dallas, New York, and Madrid, have started massive tree planting efforts to plant 1 million or more trees in the next 20 to 30 years. Most are being planted where space is available, others have created space. In Milan, the Bosco Verticale (Vertical Forest) is a building covered with trees on balconies designed to accommodate green infrastructure (Peters 2019). Many cities have adopted the use of incentives to plant trees and install green roofs. Since 1990 the Sacramento Municipal Utility District (SMUD) has partnered with the Sacramento Tree Foundation to give more than 350,000 free shade trees to residents, and since 2003 the Pennsylvania DEP Energy Harvest Program has been providing grants for energy saving projects; in 2007 more than $500,000 went for green roof projects. The Houston Downtown Management District gives matching grants for green wall projects on parking garages and buildings with blank walls (USEPA 2008b, 4–5).

Other strategies include policies and regulations to mitigate urban heat islands, including building, parking lot, and landscape policies. Examples include Tucson, Arizona, where all air-conditioned city facilities are required to use cool roofing materials. The International Council for Local Environmental Initiatives (ICLEI), which runs the Urban Heat Island Initiative Program, provides assistance to local governments with policy language and frameworks for heat island resolutions. They also coordinate workshops with local governments throughout the United States to help understand UHI impacts and mitigation strategies. Some communities use their comprehensive plans and design guidelines to put forth their policies, goals, and objectives for UHI mitigation. For example, the "Environmental Planning Element" in the Gilbert, Arizona General Plan includes mitigating heat islands as a core goal, listing policies such

FIGURE 7.3. Paris, France: green wall on the Museé du Quai Branly by Patrick Blanc. This green wall is one of several in Paris designed to be an iconic landmark as well as provide benefits of city vegetation.

as identifying projects that contribute to UHI and evaluating mitigation techniques. They partner with education institutions, companies, and government offices to promote UHI awareness, the use of cool materials and construction techniques, and the use of engineered green spaces for cooling (USEPA 2008b, 9–10, 13).

Citizen Science Program: Mapping Urban Heat Islands

The Climate Program Office (CPO), affiliated with NOAA, ran a citizen science program with eight cities in summer 2019 (with more planned for 2020) to map urban heat islands. The project is supported by the National Integrated Heat Health Information System and CPO's Communication, Education, and Engagement Division. The cities were selected based on readiness to run the campaigns and willingness to use the heat maps to build community resilience to extreme heat. Project participants drive preselected routes three times a day to collect temperature and GPS locations every second. High-resolution maps are produced to develop resilience-building actions (Dahlman 2019).

Case Study: Heat Island Maps for Richmond, Virginia

Residents in Richmond, Virginia, have experienced an increase in number of days hotter than 95°F for several years, with an increase in heat-related emergency room visits and deaths. In seeking solutions to the problem, the city collaborated with a team of scientists from the Science Museum of Virginia in Richmond, professors from Portland State University in Oregon, and citizen scientists, including students from the University of Richmond and Virginia Commonwealth University; the Virginia Academy of Science; the City of Richmond's Sustainability Office; and Groundwork RVA, a Richmond nonprofit organization that engages young people with social and environmental issues in their community. Volunteers rode bikes or drove cars three times a day to collect and map temperature points on predetermined routes. The data were entered into an open-source software package to show a detailed map of heat across the city, with differences of up to 16°F across neighborhoods. The warmest neighborhoods were those with the fewest trees, most impermeable surfaces, and highest number of heat-related calls for emergency responders. Many also happened to be neighborhoods below the poverty line, where few had air conditioning. Using a range of factors, project organizers created an index to identify neighborhoods at risk and developed action plans such as tree planting, building cooling centers and shade structures, and long-term plans to change building construction requirements. Groundwork RVA Green Team members plan to help spread the word in vulnerable neighborhoods so community members can develop personal heat adaptation plans. Members participated in a project called "Throwing Shade in RVA."

Using small-scale home models from the Franklin Institute, they used heat sensors and FLIR cameras to measure and visualize heat differences between different surface colors and then designed a variety of shade structures to cool building facades. The project was partially funded by NOAA's Office of Education Environmental Literacy Program, and the Franklin Institute's Ready Row Home exhibit is provided through the Climate & Urban Systems Partnership, funded by the National Science Foundation. The ultimate goal of the project is to develop community-scale adaptations where most of the human activity takes place. Data from the project are also being used to update Richmond 300, the citywide master plan, RVA Green 2025, their greenhouse gas emissions reduction plan, and RVAH2O, the watershed management program to reduce stormwater runoff. This urban heat vulnerability assessment can be replicated in other cities using data from the National Land Cover Database and the U.S. Census Bureau's American Community Survey (Hoffman 2018).

8

Urban Biodiversity

Variety and variability of life on earth is the simple and overarching concept of biodiversity. While this concept provides context for the study of biodiversity, historically, ecologists have used a more quantifiable definition for their work, describing biodiversity as the number and diversity of species in a community or, alternatively, the relative abundance of each species in relation to the total number (Faith 2016; Quigley 2011, 85). However, counting species numbers is being challenged with new concepts about a more complex view of biodiversity; that the health of an ecosystem depends not only on numbers but also on the diversity of their traits, or functional-trait ecology. Basically, it means that the different characteristics of species and the things they do are important to keep an ecosystem healthy and resilient. Despite agreement on the importance of functional traits, no clear definition of a trait exists, and questions remain about ranking the importance of traits. The significance of this new definition is important for two reasons: (1) it has refocused decisions about what to protect toward previously understudied areas where urban development is spreading, and (2) describing biodiversity as a functional trait is a way to convey the importance of ecosystems to policy makers and economists who might better understand how the loss of a function, rather than loss of a species, impacts an ecosystem (Cernansky 2017, 23–24).

Included in this new thinking about biodiversity is the addition of urban biodiversity as a valid field of study. Urban biodiversity focuses on novel (human-made) ecosystems and plant collections such as selected native plant species in urban green spaces (fig. 8.1). Until recently, urban biodiversity has not been a substantial part of biodiversity research, partly because of the mistaken notion that cities inherently lack biodiversity. However, more recent research shows that cities can support diversity, including threatened and endangered species, making them important for conservation. The major issue for urban biodiversity today is how

FIGURE 8.1. Perth, Australia: planted native vegetation at Chevron Parkland. The vegetation in the Parkland is designed to mimic native plant communities.

conservation can be integrated with other needs of society, such as contact with nature, for human well-being (Faith 2016; Nilon et al. 2017, 332).

Biodiversity as a discipline is rather fluid, and efforts are underway to develop a more defined research and policy agenda. The subject of biodiversity is typically included in conservation biology, which integrates many disciplines in which the goal of scientific inquiry is the protection and management of biodiversity. Scientists in traditional fields such as wildlife management, agriculture, forestry, and fisheries have joined others in ecology, restoration biology, anthropology, and economics to answer questions that can be applied to resource management. Conservation biology has been described as a "mission-oriented crisis discipline" in which the main goals are to evaluate human impacts on biological diversity and develop practical approaches to prevent extinction of species. One goal is to put an economic value on biodiversity to bring the monetary aspect to policy planning (Gerber 2010; Pennisi 1991). One of the challenges of understanding urban biodiversity is the fact that information on urban species, especially wildlife, is not easy to find. The National Center for Ecological Analysis and Synthesis (NCEAS), based at the University of California at Santa Barbara, has developed a database with species lists, abundance, and habitat types for urban wildlife in 156 global cities so far.

One thing they have learned about urban biodiversity is that cities have about 20% of the world's avian biodiversity, and younger cities tend to have more native birds. One objective of the project is to understand why birds tend to disappear as cities grow older. Scientists hope this knowledge will create opportunities for interventions and urban design strategies to keep birds in cities (Conniff 2014).

What: The Difference between Urban Biodiversity and Natural Biodiversity

Urban biodiversity (as opposed to natural biodiversity) refers to the variety and variability of living organisms *plus* the ecological systems that are transformed by anthropogenic (human) factors (Puppim de Oliveira et al. 2014, 461). Biodiversity in urban areas considers the impact of human activities on the optimal combinations of plants, animals, insects, and soil microbes, both terrestrial and aquatic. Nature in cities includes both the cultivated and managed areas such as parks and community gardens, residential gardens, and golf courses, and other green spaces such as vacant lots, roadsides, abandoned industrial sites, and cemeteries. Most of these landscapes are considered novel ecosystems: landscapes where urban development has changed the natural biologic communities to an ecosystem with no natural equivalent (fig. 8.2). In novel urban ecosystems the density and variety of flora and fauna are often reduced, resulting in loss of some ecosystem processes and biodiversity (Kimbrough 2016).

Some elements of nature, such as sunlight, water, soil, and a variety of organisms, remain even after construction of a city. What typically changes are the abundance and types of microorganisms and plant and wildlife species (Palmer 2012). Plant species variety and composition is an area of concern for biodiversity and conservation in cities, because the structure and composition of urban landscapes are often affected by human preferences for certain landscapes and the limitations of the urban environment. The number of new species that come into the new urban habitats and old species that recolonize often depend on how much vegetation is preserved in green spaces, the type of habitats surrounding the city, and the traffic patterns in cities. One study showed that vehicles traveling a motorway in Berlin transported 240 plant species in one year, 50% of which were non-native, and interestingly, traffic traveling out carried more seed species than traffic entering the city (Dunn and Heneghan 2011, 114).

FIGURE 8.2. Paris, France: large planters on a hillside contain a plant community with no natural equivalent.

Urban biodiversity can be very rich, but little is known about biodiversity patterns of individual cities, how city form affects biodiversity, and the diversity of designed and installed landscapes such as residential yards. Little research exists about the social dimension of biodiversity, but we do know that at the individual level, perceptions of urban nature are important and are based mostly on aesthetic appeal (Palmer 2012). The perceptions and aesthetic preferences of homeowners for certain types of vegetation, including colorful, neat looking, evergreen, and low maintenance plants, can reduce the variety of plants, which diminishes habitat for a number of other species. Because most urban residents interact with nature informally and subconsciously experience the positive effects of nature, they often pay little attention to other species that share their city. Sometimes called biodiversity blindness, this lack of awareness is usually due to lack of knowledge about the role of other species in urban ecosystem function (Palmer 2012).

Some doubt that human-designed landscapes (with mostly non-natives) contribute to ecosystem function and diversity. Although many private gardens have a high number and variety of plant species in a relatively small area, these "plant collections" don't guarantee taxonomic diversity. However, they are a good example of environmental vulnerabilities in even the most heavily planted spaces. Human-selected combinations of "urban" species can create new ecosystems, but they may not be healthy or resilient, depending on the functions provided by the different species and compatibility of functions. A species-rich area may be productive if many species have a variety of trait functions that help create redundancy of functions; however, if some functions are provided by only one or a few species, the ecological diversity is less robust and resilient (Cernansky 2017, 23). Designs based on an ecosystem score or ranking for functional services provided by each plant could create more ecological diversity, except little information exists about trait function or ecological services for individual plants. The exception is a few large trees in tropical areas that show the astounding capacity to support several hundred to several thousand beetles and other insects. Few gardens are intentionally designed for plant interactions, nutrient cycling, or wildlife food sources, but the trend is to use more native plants. The assumption is that native species adapted to a pre-urban ecosystem will be successful in a "new" version of their native habitat. However, unless soils, hydrology, and temperature are similar to their natural habitat, the native species may not be any better adapted to urban conditions than an introduced species (Quigley 2011, 86, 88).

So What: The Importance of Biodiversity in Urban Areas

Urbanization has profound effects on the ecological processes of natural areas when habitat destruction and fragmentation create human-made novel ecosystems that lack biodiversity. The result is loss of wildlife habitat and decreased ecosystem processes (Nilon et al. 2017, 332). Non-native animals in urban areas often prey on vulnerable wildlife populations and are often vectors for diseases and parasites that can destroy the native population, reducing the species number. In addition, they often compete for limited resources, further reducing habitat and species. Other more subtle impacts include changes in natural hydrology patterns that can destroy natural wetlands and riparian habitat, and overdrawing from local aquifers resulting in ground subsidence. The loss of a few species populations with urbanization can destabilize natural ecosystems and result in the loss of more utilitarian ecosystem services, including wind protection, climate control, watershed protection, and noise abatement. Cultural ecosystem services that are lost include the positive effects on health, recreation, crime reduction, and social cohesion, which results in fewer economic gains that come with a green city. In addition, biodiversity creates improved aesthetics and cultural desirability that encourage tourism, business, and stable long-term citizen populations. The economic value of biodiversity in urban green spaces includes the protection provided by biodiversity from stressors such as flooding and pollution (Nilon et al. 2017, 332).

An unfortunate trend in urbanization today shows an increase in development in regions identified as biodiversity hot spots, which are biogeographic regions where significant biodiversity is under the threat of destruction from development. To qualify as a hot spot the region must contain at least 1,500 species of vascular plants found nowhere else. Thirty-six biologically rich areas around the globe have been identified that have lost at least 70% of their original habitat, yet surprisingly, still support nearly 60% of the world's amphibian, reptile, mammal, bird, and plant species (CEPF 2019). The problem with developing in biodiversity hot spots is that habitat degradation is difficult to repair. Rewilding habitats, which is the reintroduction of plants and animals into their original habitat to increase biodiversity, is very difficult due to the complexity of ecosystem processes that are created over time. If certain species, functions, or processes are lost, scientists currently don't have the specific knowledge (or technology) needed to reproduce a system from a condition with no traces of the original.

Although managing for ecosystem services presents challenges for planning and policy, cities present opportunities for biodiversity conservation, primarily through planning for climate change mitigation. To promote biodiversity conservation policies, they should be linked to ecosystem services and the benefits and economic value to human well-being. Cities today are using several approaches to enhance biodiversity conditions, as shown by one study that assessed development plans for 40 selected cities around the globe. First, researchers developed an audit instrument using 34 action items (attributes) they identified as important to urban planning for biodiversity conservation. Using the audit, they reviewed the cities' plans to see which of the 34 action items were included and how the various city plans differed in their approaches. Action items included collecting baseline data on habitats and species within the city and establishing biodiversity goals, creating green networks and connectivity, using native species, and conserving specific habitats such as wetlands and woodlands. Other items included using education outreach and community engagement, developing monitoring and research activities, managing alien and invasive species, and using constructed habitats such as bioswales and green roofs. Quantitative targets included increasing species, increasing natural, constructed, and critical habitats, and decreasing non-native species. The ecosystem service goals in the review included (1) increasing water quality, tree cover, and food production; (2) using carbon sequestration and climate amelioration strategies; and (3) developing conservation goals for recreation and sense of place. Quantitative targets included reduction of heat island effects, reduction of water and air pollution, and increase in urban agriculture and number of trees. Planning-related action items included describing implementation strategies in the plan and specifying elements mandated by law or ordinance. The most common attribute, found in 80% of the plans, was the presence of an ecosystem services goal. The majority also mentioned enhancing biodiversity by improving quality and quantity of habitats, but few plans set quantitative targets for their planning goals. The audit is useful beyond reviewing existing plans; it can also serve as a checklist for cities that are considering developing biodiversity strategies as part of their climate change preparation strategies (Nilon et al. 2017, 335–336).

Now What: Questions about Creating Biodiversity in Urban Areas

Nearly every study on urban biodiversity mentions the scarcity of information about cities and biodiversity. Unanswered questions range from big-picture planning decisions, such as habitat size, connection, and quantity, to small-scale questions about habitat quality, diversity of traits, and native vs. non-native species. One of the biggest challenges appears to be lack of information about urban species, particularly wildlife, and how to integrate the needs of nature and the needs of society. Several projects are currently under way that are designed to fill the information gaps. For example, a study in the journal *Landscape and Urban Planning* looked at better ways of understanding urban wildlife and habitat in combination. The study used birds (due to their ease in counting) as bio-indicators for other wildlife types. The study proposes a marriage of i-Tree and eBird, two current software programs, one to record data on urban tree cover, the other for birders around the world to log bird sightings. Together the information could help researchers and planners understand which trees are good habitat and which birds are using which trees (Conniff 2014).

Several organizations are building databases of urban species to increase understanding of habitats. Plant and bird databases are the most popular because plants and birds are the easiest to count and document, especially for citizen science projects. As mentioned, NCEAS is creating a plant and bird database for a comparative approach to the study of urban biota. The project is analyzing data on plants and birds in 140 cities around the globe to help develop different planning practices for monitoring and preservation of biodiversity. The goals of the research are (1) to compile large diverse data sets of the birds and plants of cities around the world; (2) to compare the data from the urban habitats for patterns and ecological responses of birds and plants; (3) to gain an understanding of the social constraints on biodiversity in urban areas; and (4) to create recommendations for monitoring biodiversity in cities (Aronson et al. 2019). The team has discovered that human factors are more influential than region or city location. Older cities with greater proportions of intact vegetation preserve more plant species, showing the importance of remnant vegetation and restoration of natural areas. Most of the bird and plant species catalogued were native, and cities supported populations of 36 threatened bird species and 65 threatened plant species. However, cities supported 92% fewer bird species and 75% fewer native plant species than typical for similar undeveloped land (Lathrop 2014).

Two other network/database projects include the Urban Biodiversity Research Coordination Network (UrBioNet) and TRY, an open access plant database. UrBioNet is funded by the U.S. National Science Foundation and was established to develop a network of scientists, students, and practitioners interested in urban biodiversity. It is a forum for data sharing and collaboration on urban biodiversity research, design, planning, and management. Its mission is to expand databases globally to identify general patterns and processes shaping urban biodiversity and to quantify the relative importance of different factors on those patterns. Information is used to develop recommendations for monitoring biodiversity to help practitioners and policy makers in urban design and planning (Kimbrough 2016). TRY is an open-access plant trait database developed by a network of vegetation scientists headed by the Max Planck Institute for Biogeochemistry. As of March 2019, the database contained nearly 12 million trait records on about 300,000 plant species. Almost half the data is geo-referenced and covers 15,000 global measurement sites. Traits such as leaf area, wood density, growth form, seed mass, leaf longevity, and plant life span are being documented. These traits are key to understanding and predicting ecosystem adaptations needed to meet the challenges of biodiversity loss (TRY 2019).

Scientists in a variety of conservation biology disciplines are addressing questions that get to the heart of their "mission-oriented crisis discipline" by considering issues directly applicable to planning and design of green spaces in cities. To develop management and restoration plans, it is important to identify the conservation value of different types of green spaces. Five ecology-based questions applicable to planning and design of green spaces have been proposed. The first question concerns the optimal size of a green space needed for biodiversity conservation. The amount of green space is an important determinant of biodiversity; however, what is unknown are the sizes needed for different species and how the surrounding urban intensity and structure impact the green space. The second question investigates which factors, such as size or number, limit population size. For example, mobile species may need more than one green space, while less mobile species might find all their resources in a single patch. Question 3 asks how green space heterogeneity (variability in structure and distribution) affects plant and animal combinations. Little is known about how heterogeneity affects species richness and viability, but this information is critical to plant selection for the design of green spaces. Question 4 relates to connectivity of green spaces and addresses the merit of corridors versus stepping-stones. Most urban plans promote corridors; however, mobile species may do better with closely spaced patches because they offer multiple pathways (path redundancy) through a given area. The last question is about green spaces as ecological traps and population sinks. Green spaces operate as ecological traps when they limit reproduction and survival rates, even though they are first-choice habitats for some animals. Population sinks are last-choice habitats after better quality green spaces are filled, and they also limit population growth. The problem is that scientists know little about how and why these spaces are selected or avoided by animals or about their conservation value. These questions have implications for the typical two-prong approach to conserving biodiversity, which includes preserving remnant natural habitats and increasing green infrastructure networks, such as green corridors. Although private gardens and pocket parks are often dismissed as low-quality habitat, they may be important stepping-stone microhabitats for a large variety of mobile species (Lepczyk et al. 2017, 802–805; Nilon et al. 2017, 332). The study of urban biodiversity is a young field in the larger discipline of conservation biology. Although many questions remain to be answered, conservation biologists agree that not only should cities consider conservation goals, but it is also critical that they do so to protect cities from the uncertainties of climate change.

Citizen Science: Monarchs and Milkweeds Program

The University of Wyoming Biodiversity Institute has developed a variety of citizen science programs. One example is the Monarchs and Milkweeds program to gather data on where, when, and how many monarchs and milkweed plants are found in the state. The program goal is to find out where monarchs migrate through Wyoming, at what time of the year they migrate, and how many are moving along the migration path. The other goal is to learn where and which species of milkweeds exist in Wyoming to help develop more milkweed habitats for migration routes by harvesting some of the seeds (The Biodiversity Institute n.d.).

Case Study: Knoxville's Urban Wilderness

Knoxville Urban Wilderness is a recreational, cultural, and historic preservation initiative championed by Legacy Parks Foundation. This remarkable outdoor adventure area incorporates 1,000 acres of forested land along the south downtown Tennessee River waterfront. Visitors can hike, bike, climb, paddle, or explore the woods—all within the heart of the city. With more than 50 miles of trails that connect parks and historic sites, it creates an outstanding wildlife, recreation, and cultural corridor. Included in the wilderness area are lakes, bike and walking trails, historic sites, and a nature center. The Legacy Parks Foundation offers the opportunity for citizens to participate by becoming a Friend of Legacy Parks. The wildlife management area is known for its abundance of wildlife, especially songbirds in the 315-acre wildlife sanctuary close to downtown. Community support is high for the Legacy Parks Foundation, which has more than 25 business and organization sponsors, including 17 bicycle and running/walking clubs. Road rides, fun runs, outside yoga, fishing tournaments, and art-in-the-park are just a few of the 300 events a month in the summer. The project has been very successful from an ecological, educational, and economic standpoint (Kimbrough 2016).

9

Urban Ecosystem Services

Ecosystem services (ES) is a complex concept that attempts to describe the human/nature relationship in terms of benefits and value to humans. To understand the concept of ecosystem services, it is helpful to remember the description of a natural ecosystem as a collection of living plants, animals, and microorganisms in a specific place that interact with the non-living environment such as the atmosphere, soil, and sun to create an interdependent natural system (DEWHA 2009). Urban ecosystems are different because they include humans, who often modify nature to meet society's needs in ways that can be harmful or beneficial to the natural ecosystem. As a result, urban areas have their own ecosystem classification—the human ecosystem—and ecosystem services in cities are often referred to as Cultural Ecosystem Services (CES). CES takes into account human/social and nature/ecology system dynamics in urban areas, considering ecology and technology, economics, politics, and history. Although cities are often linked to ecocide (the destruction of entire ecosystems by humans), previous chapters have shown that urban areas can support a healthy and dynamic human-dominated ecosystem (Pritzlaff 2019).

The study of urban ecosystem services is broad and challenging and includes every specialty in the life sciences, including plant, animal, soil, and water sciences, and socioenvironmental sciences such as public health, social sciences, anthropology, and urban planning. Research in this field is critical to inform all levels of environmental policy, including economic strategies, sustainable cities development, food and medicine programs, and management and conservation plans. Historical patterns of early urbanization show that cities grew by transforming the environment to increase benefits (such as survival odds) and reduce risks, even if it meant future environmental degradation. Today, in the era of the Anthropocene, we need to rethink our relationship with nature. If we want improved human well-being and reduced environmental risks, we will need new visions for ecology in urban areas that incorporate nature and social capital (Elmqvist, Redman et al. 2013, 20–21, 25). For example, Payment for Ecosystem Service (PES), also known as payments for environmental benefits, is a relatively new idea that puts a value on an ecosystem service by offering incentives to landowners to manage their land to provide some sort of ecological service, such as water purification. The most common example is putting private land, such as livestock ranches, in conservation trusts that are managed to benefit others (Conservation Gateway n.d.).

Few citizen science (CS) programs directly assess ecosystem services. Projects are challenging because the complexity of ecosystem services necessitates the use of complicated protocols. Volunteer training can be difficult, leading to potential sources of error, especially when participants lack skills in species identification. However, CS can facilitate the collection of large amounts of data over broad temporal and spatial ranges with low, or no, staff costs (Schröter et al. 2017, 87).

What: Defining Ecosystem Services

Ecosystem services are the benefits humans derive from the natural environment to improve their health and well-being. In 2005 the Millennium Ecosystem Assessment project identified specific ecosystem services in four categories: (1) provisioning—products obtained from nature, such as food (fig. 9.1), fiber (fig. 9.2), fuel, and drugs; (2) regulating—activities performed by nature, such as pollination, seed dispersal, and flood and disease control; (3) cultural—intrinsic benefits such as spiritual enrichment, health, recreation, social relations, and aesthetic experience, and (4) supporting—the basic ecological functions such as production of oxygen, soil formation, nutrient cycling, and habitat creation (Millennium Ecosystem Assessment 2005, 7; DEWHA 2009, 8; Green Facts n.d). To understand how ecosystem services contribute to human well-being it's important to understand what constitutes well-being. All humans have the desire for security and health, access to basic materials for a good life, good social relations, and freedom of choice and action. We gain our basic materials and good health through the provisioning services that provide food, water, fuel, and shelter. Good health and security also come from the regulating services that provide climate, flood, disaster, and disease control. Cultural services support good social relations and health and security by providing the spiritual, aesthetic, and educational opportunities that create social cohesion. When ecosystem services are readily available, other

FIGURE 9.1. Salerno, Italy: street vendors sell local food from a food cart on the Salerno waterfront.

FIGURE 9.2. Izmir, Turkey: silk pods soaking in water to loosen threads for weaving into silk rugs.

environmental factors such as economic, technological, and cultural factors are more prevalent, creating conditions more favorable for freedom of choice and opportunities (Millennium Ecosystem Assessment 2005, 50).

Two conditions—biodiversity and resiliency—are the cornerstone of ecosystem services. Biodiversity drives ecosystem function, and loss of biodiversity directly influences the ability of an ecosystem to produce services. Resilience is the capacity of the system to absorb impacts and disturbance (often human-made) and continue to function and provide services. Although ecosystems have varying degrees of stability, most are resilient until the rate and scale of disturbance are so great, such as building a city, that the essential processes change (DEWHA 2009, 5–6).

To investigate the consequences of ecosystem change for humans, the Millennium Ecosystem Assessment project was proposed in 1998 at a meeting held at the World Resources Institute. In 2001, United Nations acting Secretary General Kofi Annan featured it as one of five major initiatives for "Sustaining our Future" in his Millennium Report to the United Nations General Assembly. To prepare the assessment, 1,360 experts from around the globe came together to provide a state-of-the-art scientific appraisal of the condition and trends of the world's ecosystems and the services they provide, as well as the scientific basis for action to conserve and use them sustainably. The focus was on how humans have changed ecosystems, how the changes have affected humans, how they will affect them in the future, and what types of responses can be adopted at local and global scales to improve management. Some of the important conclusions were that (1) the human destruction of some ecosystems has resulted in substantial and largely irreversible loss of diversity; (2) the changes have contributed to substantial gains for human well-being and economic development—however, at a great cost to the degradation of many ecosystem services and increased poverty for some people; (3) the degradation could grow significantly worse in this century; and (4) the challenges of reversing the degradation while still providing services will involve changes in policies, institutions, and practices. The assessment was unique at the time because it focused on ecosystem services and the link to human well-being. Some of the emergent findings included: (1) 60% of the 24 ecosystem services examined are being degraded; (2) the likelihood of abrupt changes is greater, such as dead zones, collapse of fisheries, and disease emergence; (3) more priority should be given to drylands (as opposed to tropical forests and coral reefs) where the ecosystems are most fragile and human populations are growing rapidly; and (4) the issue of excessive nutrient loading of ecosystems is not receiving enough policy attention (Overview of the Millennium Ecosystem Assessment n.d.).

So What: The Importance of Ecosystem Services in Urban Areas

From a policy perspective, sustainable development should include ecosystem services in the decision-making process. The distribution and type of ES provisions depend on two things, the number, size, and distribution of green and blue (water) areas, and a distribution of population and urban functions that establishes who benefits the most from ES (Cortinovis and Geneletti 2018, 298). To address inequities, it is important to put a value on ecosystem services. Most of the time we are not aware of ES because they are not easy to observe and identify—until they cease to exist. For example, a river provides many services, including fresh water, food, transportation, beauty, and recreation (fig. 9.3), until it becomes polluted. Then the value of these lost services becomes clear, especially considering the substantial cost to restore the degraded system (DEWHA 2009, 2). If we don't put a value on ES—that is, we regard them as "free"—no incentive exists to value, protect, or restore them. The most highly valued services are those that can be measured and are accessible. All four categories of ES

FIGURE 9.3. Paris, France: citizens enjoy the water and trees along the banks of the Seine River.

can be included in market economics. For example, provisioning (mostly food and fiber) has economic value, including the cost of support activities such as fertilizer and pest control to maintain the services. Regulating services, such as seed dispersal, and pest and disease control can be artificially supplied and thus assigned a value. Other services, such as pollination, can be assessed on the extent to which crops produce harvest. Some services, such as carbon sequestration and climate control, have been outside the market, but more recent evidence is sufficient to price and integrate them into the market. Although supporting services are traditionally unvalued, soil and biodiversity conservation are now acknowledged, usually through government investments to protect and improve them. Cultural services, such as indigenous lands, clearly have value, but they are not normally priced or included in markets. However, there is opportunity for nature-based tourism and recreation to be valued based on visitor numbers and expenditures on outdoor activities. Other values such as aesthetics and sense of place can use surrogate measures, such as increase in real estate prices for properties with visually pleasing landscapes. An effort to value all ES will generate more investment to ensure the long-term supply of services; conversely, undervaluing the full range of ES can result in species extinction, air and water pollution, poor soil health, and associated impacts on economy and well-being (DEWHA 2009, 10–11, 16–18).

Urbanization often changes people's perspective of nature and the natural environment. For most people "nature" is the managed green spaces, such as parks, sports fields, and yards in urban areas (Dickson and Hobbs 2017, 180). Many studies have shown that people derive both physical and mental health benefits from green spaces, and they are usually some of the most popular and beloved spaces in cities. Of particular importance to human well-being in urban areas are the cultural ecosystem services, such as spiritual and artistic enrichment, mental health, recreation and leisure opportunities, social relations, and aesthetic experience that people derive from the environment. These services represent very familiar and personal experiences with nature in the urban environment which could be a motivating factor for willingness to conserve natural areas in cities (Dickson and Hobbs 2017, 180). Cultural services are the most difficult to value because they are intangible, and many can't be measured; however, some cultural phenomena are tangible and can have social and scientific value. For example, spiritual enrichment might come from preserving ancient burial grounds, sacred sites, or archaeological sites. Ecosystems with

FIGURE 9.4. Perth, Australia: multimodal path through parklands along the Swan River. The parklands and river are cultural and natural ecosystems that provide recreation opportunities.

unusual geologic formations or rare and endangered plant or animal species are also tangible and have scientific value.

One characteristic of CES is human involvement in the production of cultural ecosystems. Green spaces created through human perception of nature usually focus on nature that is most aesthetically pleasing and functional for leisure and recreational activities (fig. 9.4), in other words, nature specifically designed to provide cultural services. Another feature of CES is the relationship to place; for example, in recreation and leisure activities, the setting creates the experience for the user, and each place will create a unique experience. Human-environment co-production is perhaps what most distinguishes CES from other ecosystem services (Dickson and Hobbs 2017, 183, 188). The use of turfgrass in urban green spaces is the most ubiquitous example of nature designed to provide cultural services. It is also the most controversial, primarily because of the perceived lack of ecological ecosystem services from turf. Turfgrass does provide several ecosystem services, but typically not to the extent of other urban vegetation. For example, turf moderately decreases urban temperature compared with absence of vegetation and has positive carbon sequestration, but only because of higher accumulation in the soil. Compared with other vegetation, turf can remove more nitrogen and phosphorus from stormwater by increasing water infiltration, which increases groundwater recharge and helps prevent erosion. While soil in turfgrass supports a healthy microbiome, including insects, it provides little food or shelter for other species, the exception being birds that feed on insects. Improved landscape aesthetics with the use of turf is the most significant CES for homeowners who cite increased aesthetics, higher property value, and more recreation space as positive benefits that increase the perception of improved quality of life (Monteiro 2017, 151–156).

Now What: New Strategies to Improve Ecosystem Services

Investing in ecosystems for sustainable development is a new approach in urban areas, which is a shift from destroying or harming ecosystems during development. Strategies include public-private investment, collective arrangements for cooperative action, and market-based approaches. Policy tools include persuasion (such as behavior change and best practices), regulations, and financial incentives or disincentives, such as tax rebates or restrictions. Market-based policy tools include incentives that influence behavior by creating markets; these can include tenders, offsets, cap and trade schemes, and eco-certification programs (DEWHA 2009, 20). A study on the inclusion of ES in urban planning documents revealed a high number of strategies in use to address urban ES and also a variety of methods for implementation. However, it was also noted that planning practices only partly relied on scientific knowledge, and little guidance was provided on how to incorporate ES information in the planning process. Plans typically contain no analysis of the demand for ES or who will benefit, plus few methods exist to assess urban ES that consider the multifunctions of ecosystems at different scales (Cortinovis and Geneletti 2018, 306).

To address the issue of incorporating ES information into planning, a document from the MEA, "Ecosystems and Human Well-being: A Framework for Assessment," provides some useful information for decision-making. Suggestions include (1) identifying and evaluating policy and management options for sustaining ecosystem services and matching them with human needs; (2) identifying and classifying the benefits to people and communicating the benefits in language that can be applied to urban development issues; and (3) identifying and providing answers to the social and ecological questions to improve management of ecosystems (DEWHA 2009, 8–9; Millennium Ecosystem Assessment 2003). Putting a value on the benefits of green spaces is also important to deter development pressure in urban areas. To this end, a five-step framework was developed for assessing green space contribution to ES in urban areas. The purpose is to evaluate the effects of potential land use changes and proposed development plans, and to identify where ES is lacking in the planning process. A simple method was developed to estimate total effect and value of ES on a site by using a limited number of indicators of functional traits and allowing for separation of the ES contributed by different components such as trees and shrubs. Steps include: (1) compiling an inventory of indicators, including leaf area index, leaf area density, canopy cover, presence of bees, diversity of songbirds, number of plant species, and percent permeable surface; (2) applying factors to rate the effectiveness of each indicator by abundance and diversity; (3) estimating the effects, whether weak, moderate, or strong; (4) estimating the benefits of each ES as the perceived value by users; and (5) estimating the total ES value of the ecosystem as the sum of benefits of all the considered ecosystem services (Andersson-Sköld et al. 2018, 274, 276, 278–280).

Citizen Science Program: Earth Challenge 2020

Earth Day Network, the Woodrow Wilson International Center for Scholars, and the U.S. Department of State's Eco-Capitals Forum launched Earth Challenge 2020 as the world's largest coordinated citizen science campaign in honor of the 50th anniversary of Earth Day on April 22, 2020. The challenge is for millions of global citizens to collect more than a billion data points on air quality, water quality, biodiversity, pollution, and human health. The hope is to leverage information to inspire collaborative action and influence policy decisions (Earth Challenge 2020 n.d.).

Case Study: New York—PlaNYC

New York City is one of several case studies for assessment of ecosystem services presented in *Urbanization, Biodiversity and Ecosystem Services: Challenges and Opportunities, a Global Assessment*. Chapter 19, *Local Assessment of New York City: Biodiversity, Green Spaces, and Ecosystem Services*, describes the goals set out by NYC in its environmental and economic sustainability plan: PlaNYC. New York, the world's first megacity, has a diverse population of more than 20 million people yet supports a rich biodiversity in its estuarine and terrestrial ecosystems, with more parkland than any U.S. city. It is one of the greenest cities in the United States, with a relatively small ecological footprint due to high public transit use and multifamily housing. The landscape has been extensively altered; in 1609 the watershed in which it exists was almost entirely forested; by 1880 about 70% had been converted to farmland and the rivers were widely dammed for agriculture, power, and drinking water. By the mid-1880s NYC was an early leader in park development with Central Park and Prospect Park, and by the 1950s a sustainability vision was created for a healthier and cleaner city based on the writings of several pioneering urban planners, including William Whyte, Jane Jacobs, and Ian McHarg, who used NYC as their living laboratory. Today there is the Mayor's comprehensive PlaNYC 2030, launched in 2007, that articulates 132 specific environmental initiatives, including the MillionTreesNYC campaign, the NYC Green Infrastructure Plan, and a waterfront revitalization program. Assessments by the city government, NGOs, and several universities reveal the broad number of ecosystem services provided by the green spaces; however, they also highlight the significant challenges in protecting and improving their biodiversity, including pollution, sea level rise, stormwater management, climate change, and increased population. To address these challenges, NYC has committed to improving the environmental quality of its urban green spaces to increase ecosystem services with three key actions: (1) commit to acquiring more data to facilitate informed decision-making; (2) create more coherent governmental support with long-term planning documents such as the PlaNYC and the NYC Green Infrastructure Plan; and (3) enlist the help of organizations and civic engagement to invest in natural areas and green infrastructure (McPhearson et al. 2013, 356–357, 370–371, 376).

PART II

ECOLOGY *OF* CITIES

Bionetworks

Part II is concerned with the bionetworks of cities, including green infrastructure, novel ecosystems, and built structures that create the city ecosystem. Chapters 10 through 14 describe the bionetworks that enable nature and the built environment to function as a system. Thinking of city ecosystems as a network makes it easier to understand the concept of material flows into, around, and out of the city. Bionetworks, short for biological networks, are the natural areas and green spaces (sometimes called nodes) located throughout the city and connected by linear corridors, such as road verges, riparian shorelines of rivers, and planted utility easements, to form a net-like pattern to create the bionetwork. Green spaces can include intentional (human-made or novel) ecosystems, such as constructed wetlands, stormwater ponds, and pollinator gardens, or unintentional habitats such as abandoned lots. Other green spaces include recreation areas such as parks, golf courses, and playgrounds. Urban structures, including canals, plazas, buildings with green roofs, and stormwater management facilities, are also considered nodes in the network. The bionetwork, often referred to as the green infrastructure of the city (chapter 11), provides numerous ecosystem services, including flood and temperature control, soil stability, and resource protection. Many cities are using green infrastructure principles and strategies as part of their resiliency plans to future-proof their city in flood-prone areas and to mitigate sea level rise from climate change. The numerous connections in the network facilitate a high number of interactions between social and ecological structures and functions that influence ecosystem health in urban environments.

Urban ecosystems and metabolism are sometimes compared to the "eco-system" of the human body; humans require energy and material flows from their surroundings, such as food, water, and air, and they transport and utilize the inputs throughout the body and move wastes, such as heat, solids, and CO_2 from their system. In the same way that humans are not separate from their surroundings, urban areas are also connected to the land around them and require energy and material flows. They can have far-reaching impacts on the "city-region" in which they are located as they consume energy and eliminate wastes, such as air pollution, heat, waste-water, and industrial waste that flow in and out of the city. The ecological footprint of a city can also extend hundreds of miles away to agricultural lands that produce food, and to forests and mining areas that supply mate-rials for construction in the city (Douglas and James 2015e, 58–61).

The interactions that occur between social and ecological systems through bionetworks provide the basis for an ecosystem approach to stew-ardship (care) of natural resources in cities for more sustainability. The ecosystem approach is applicable to urban ecology because it is a strategy based on biodiversity and the recognition that humans are the managers of biodiversity in urban areas. The approach supports integration of land, water, and natural resource management for conservation of biological diversity, sustainable use of resources, and fair and equitable sharing of benefits (ecosystem services), which are three objectives of the 1993 Con-vention on Biological Diversity (CBD). The CBD, an international, legally binding treaty developed by the United Nations Environment Programme (UNEP), is regarded as the key international instrument for sustainable development. The twelve principles of the ecosystem approach are the foundation for actions outlined in the CBD, which operates under the pre-cautionary principle—that is, when there is a significant threat to loss of diversity, precautionary measures should be taken to minimize or avoid the threat even if some cause-and-effect relationships are not fully scientifically established. The CBD also acknowledges that conserving biological diver-sity will take substantial investments; however, it notes that conservation is necessary for the significant economic, environmental, and social benefits humans receive (UN Environment Programme n.d.).

A variation of the ecosystem approach to sustainability, the socioecolog-ical system (SES), uses a policy approach to management, outlining prin-ciples for management activities that are applicable to policy development. The principles are complementary to those in the ecosystem approach, with some overlap, but they provide direction for policy that supports the ecosystem approach. SES principles, summarized here, are stated as actions that can be directly written in codes and policy. SES recommends maintaining diversity and redundancy, ensuring connectivity with networks, and managing over different time scales to observe slow changes. Flexibility of management is also important as socioecological systems are complex and adapt over time or with unusual situations. Experimentation and trial efforts are encouraged to acquire new knowledge. Using both the ecosystem approach and the socioecological system can help urban ecosystems and cities transition to a more sustainable state. Examples of shifts toward greater sustainability that are happening today include use of green energy sources, such as wind and solar energy, using new materials that generate power, using vegetation to save energy for more efficient buildings, and using improved and innovative green infrastructure based on ecology (Douglas and James 2015e, 68–71).

10

Novel Ecosystems

Novel ecosystems or landscapes (sometimes called non-analog, emergent, adaptive, or hybrid ecosystems) is an environmental concept with a history in conservation biology and restoration ecology. Although the idea of a novel ecosystem seems fairly straightforward—basically a human-built landscape with no analog (similar) landscape in nature—the concept has been complicated by competing views from various ecology experts with disagreements over the usefulness, accuracy, and application of the term. The concept of a novel ecosystem grew from the acknowledgment that many ecosystems around the globe were rapidly transforming as a result of biotic (biological) changes and abiotic (physical) changes, or a combination of both, and from lack of a specific category of ecosystem to describe the new environmental conditions (Hobbs, Higgs, and Harris 2009). Over time, a variety of evolving definitions tried to capture the increasing complexity of application to different ecosystems. Lack of a definition made it difficult to formalize the concept as a useful statement, creating debate among restoration ecologists, conservationists, and urban planners with different goals and objectives for conservation efforts. The debate centers on different opinions of what to do with places that have been greatly changed by people. Restoration ecologists have always sought to bring ecosystems back to their historic condition; however, others say some ecosystems have changed so much they cannot be restored and should instead be managed for ecosystem services. This "new conservation" approach is unacceptable as a restoration practice for restoration ecologists. At the same time, conservation biologists, who focus on protecting unchanged places, also took issue with the new conservation concept, saying the goal was no better than trying to restore ecosystems to a historic condition, advocating instead for "old conservation" by limiting human interference with new protection strategies. Currently the debates continue about such issues as defining novel ecosystems—that is, novel in what aspect and relative to what, the degrees of novelty (degree of dissimilarity relative to a baseline), and the distinction between change and novelty, noting that landscapes change, but the result is not always a novel ecosystem (Novel Ecosystems 2016). While the debate continues between restoration ecologists and conservationists, those who work in the urban sphere have generally accepted that almost all landscapes in cities differ from anything historical; that is, they are novel. Meaning they are usually not self-sustaining (with the exception of some vacant lots or roadsides), are almost always managed by humans to some degree, and could not persist without them, but they still provide ecosystem services, and sometimes cultural services.

The concept of novel ecosystems as they relate to urban conservation practices revolves around four key issues: (1) the ability to identify the historic state; (2) identifying irreversible thresholds that prevent restoration to the historic state; (3) determining the impact of non-native species and restoration activities on natives; and (4) the concept of a hybrid state that includes elements of novel and historic ecosystems (Miller 2016). To describe novel ecosystems in urban areas, the definition proposed by Nathanial Morse and others seems the most appropriate: "A novel ecosystem is a unique assemblage of biota and environmental conditions that is the direct result of intentional or unintentional alteration by humans, sufficient to cross an ecological threshold that facilitates a new ecosystem trajectory" (Morse et al. 2014). How urban scientists interpret and apply this definition as a basis for identification, creation, and management of urban green spaces depends on the ecological characteristics of the landscape, the social and cultural context, that is, use of the space, and the intent behind policy development and conservation strategies for the space.

What: Characteristics of Novel Ecosystems

Urban ecologists, urban conservationists, and land planners generally use a simpler, more pragmatic and applied description: "human-built ecosystems that have been altered in structure and function," as a way to define novel systems. This description implies changes in qualifiers (such as species composition and structure) by noting that structure and function have been altered, allowing for broader application to many urban situations to identify and create landscapes that compare to natural analogs for selected functions. While urban ecologists agree that novel urban ecosystems may not look like their natural counterparts, they maintain that a designed hybrid ecosystem can function in a similar capacity. Hybrid systems are

FIGURE 10.1. Gainesville, Florida: recently planted pond edge on University of Florida campus. A combination of trees and shrubs that can withstand periodic flooding was used to create habitat and add diversity to the shoreline.

described as having some original characteristics and some novel elements which presumably would make them easier to restore (Hobbs, Higgs, and Harris 2009). Definitions aside, four criteria are useful to help identify, create, and manage a novel ecosystem: (1) human agency, intentional or unintentional, creates novel ecosystems (fig. 10.1); (2) the system has crossed one or more thresholds, a point at which the change in ecosystem properties are difficult or impossible to reverse; (3) the system consists of new combinations of species and relative abundance that make it unique from all other ecosystems in the same biome, for example, the presence of non-native or invasive species; and (4) the abiotic and biotic properties are self-sustaining and able to persist without, or in spite of, human intervention (Morse et al. 2014). This means opportunities exist for humans to improve the health and functionality of a degraded urban environment through the use of novel ecosystems that don't have the same constraints as traditional restoration and conservation practices.

Most novel landscapes designed for ecological function in the city target a very specific ecosystem service or function, such as clean water, carbon sequestration, or wildlife habitat, and adapt the design to fit in the urban framework. This targeted concept acknowledges the fact that novel

FIGURE 10.2. Tampa, Florida: installing shoreline buffer plants in a stormwater pond to provide habitat and buffer wave action to reduce shoreline erosion.

ecosystems in cities are here to stay and the near impossibility of reverting to the original state of nature. The question for urban ecologists is to what extent managed biodiversity can provide the greatest number of ecosystem services and establish self-sustaining populations. A socio-ecosystem approach to assess and enhance urban biodiversity with hybrid ecosystems is more likely to succeed. This approach takes into account the interactions and functioning of both environmental and social issues and acknowledges that cities have the potential to host a great variety of species. Examples of intentional novel ecosystems include created wetlands and aquatic habitats (fig. 10.2), biodiversity corridors, wildlife corridors, butterfly and pollinator gardens, and wildlife gardens. Examples of unintentional novel ecosystems include vacant abandoned lots, naturally vegetated stormwater channels, and roadside ditches.

Novel ecosystems are more likely to be found in cities because of the great variety of conditions that can support different species, including a high number of introduced species. Those who support the novel ecosystem idea also take a more pragmatic approach to the use of non-native species, stating that management decisions should be based on a species' impact—not on their origin—and that urban areas usually support a mixture of native and non-native species in a functioning species assembly

(Mitra 2017). Urban novel ecosystems must respond and rapidly adapt to a variety of urban conditions, including disturbed soils, altered hydrology, temperature extremes, and interaction with non-native species and loss of coevolved species. The most successful novel ecosystems in some situations, such as abandoned lots, are those where organisms can self-assemble and adapt to each other through certain mechanisms, including behavioral, physiological, and genetic changes (ESA Annual Meeting 2016). It is also agreed that within cities the crucial factors that determine the biological richness of novel landscapes are the intensity of surrounding urban land use, habitat continuity (which may require more preservation actions), and the type of management practices used on green spaces (Kowarik 2011).

Novel ecosystems are everywhere in urban areas and are familiar to city dwellers. Although little is known about their potential to improve biodiversity and sustainability, they do have the potential to bring nature to people in urban communities who may never experience nature outside the urban context. For this reason alone, novel landscapes in urban areas should be embraced and encouraged. Some urban ecologists refer to different types of novel ecosystems within cities based on the history of the location, the level of transformation, and the type of novel ecosystem that was intentionally or unintentionally created. This line of thinking is labeled the "four natures approach," as follows:

Nature of the first kind (also called pristine) are areas of low transformation, including remnants of natural ecosystems such as forests or wetlands.

Nature of the second kind (also called agricultural) are areas of medium transformation, including remnants from other earlier habitat changes such as old agricultural fields.

Nature of the third kind (also called horticultural) are medium to highly transformed areas and include green space established after habitat destruction, such as parks (fig. 10.3), yards, and gardens.

Nature of the fourth kind (also called urban-industrial) are highly transformed areas, including ecosystems that developed in vacant lots or transportation corridors and industrial sites.

The purpose of the categories is to guide conservation efforts. Knowledge of the historical land use and the level of change that took place on the site can hold keys to the soil quality, original hydrology, and original versus introduced species—information that is helpful when trying to create some

FIGURE 10.3. Perth, Australia: native plantings in Chevron Parkland create a nature-based playground that includes rocks and tree stumps for climbing.

habitat continuity for some of the original species assemblages. Recent research has shown that urban areas can contribute significantly to biodiversity conservation if urban habitat types are preserved or created across the entire spectrum of the four natures (Kowarik 2011, 1978–1979). Some ecologists also refer to nature of the fourth kind as "chance ecologies" or the "new wild/new nature" idea, where natural rewilding appears without noticeable human agency in abandoned lots or fields and persists untouched by humans. These bits of city wilderness can sometimes be considered undesirable habitats of low quality, as many include garden escapes and native and non-native opportunistic species, yet they may be the closest to wild nature some urban dwellers get to experience (Collier 2016).

So What: Environmental Concerns about Novel Ecosystems

Cities are created for people, and conservation efforts must meet the often-competing needs of people and nature, with the common goal of conserving and enhancing biodiversity to improve the health of the environment and the well-being of people. The conservation target for cities should also focus on creating ecosystems that maximize natural functions, concentrating on key species for conservation, but also allowing non-native species well adapted to the built environment. Prior uses and current social and cultural values and political influences must also be considered. This concept of layered landscapes is more realistic for ecological restoration in cities and maximizing biological potential of natural areas (Schaefer 2017). Biodiversity in novel landscapes can be impacted by a variety of socioeconomic factors such as income, availability of ornamental (non-native) plants, species selection for landscaping, management of urban green spaces, and movements of humans. Wealthy neighborhoods typically have more introduced, non-native species, while vacant lots in lower income neighborhoods tend to have a mix of species. While green-space management activities can reduce threats, such as insects and diseases, they also enable the persistence and spread of non-natives, as human dispersal of seeds and plants is a more significant factor in biodiversity than habitat-related factors (Kowarik 2011). See chapter 8 for further discussion on urban biodiversity.

Conservationists often argue that non-native species are harmful to the environment, pointing to several risks associated with using non-natives. Many fear that the displacement of native species is caused by overuse of non-natives. While more studies are needed to determine if a causal role exists for native decline when non-natives are introduced, it is recognized that the longer a non-native species is available on the market and used in landscapes, the more likely it will spread as a result of propagule pressure (or introduction effort). Propagule pressure is a composite measure, including the quantity, quality, and frequency of the number of individuals of a species used in a region where they are not native. Some are concerned that using high numbers of non-natives will increase biotic homogenization (when spatially separate communities become the same over time), as the same suite of plants show up repeatedly in cities. However, some historical studies over millennial time spans show considerable homogenization of natives in some cities in the past, but diversity increased at the local scale as cities grew and transportation and trade routes increased. The increase in diversity does not imply, however, a similar increase in functionality or ecosystem services. Cities are prominent points of entry for non-natives, but they can also be a point of dispersal; birds, traffic, garbage waste, and moving water are often blamed for movement of non-native species outside the city to the surrounding countryside. Impacts of non-natives on the higher trophic levels are a concern as most wild urban animal species prefer or rely on native species. Some studies, however, show that the heterogeneity (mixed composition) of the garden is more important than the status of the individual plants. Other risks include human health impacts, but few exotic species add risks, as dangerous plants are typically not bred for the consumer market. Native plants are usually responsible for more allergies or toxic reactions in most people. While overuse of non-natives can diminish some ecosystem services, positive effects have also been found, especially with cultural services, but also with some basic environmental services such as carbon sequestration (Kowarik 2011).

Now What: New Strategies for Creating Novel Ecosystems

Current conservation strategies typically focus on restraining urban growth to preserve native ecosystems, repopulating native species in urban habitats, and preserving and improving relic habitats. These are important strategies and should continue; however, they don't address the entire spectrum of urban nature, such as profoundly altered sites or highly developed horticultural sites, since their bias is toward restoration. Instead of relying on replication of historical communities as the framework for urban conservation, cities would be better served by emphasizing the macro-dynamics that are influencing colonization, distribution, and loss of species in cities (Dooling 2015, 101). Urban ecologists argue that a new paradigm is needed to shift toward considering all of urban nature, because the focus on native or original nature ignores the benefits from "other" urban nature (Kowarik 2011). The new paradigm would challenge conventional conservation in several ways. First would be the concept of "conservation value": do these sites have environmental value and, if so, how do you measure value? The worth of highly developed sites may be in their social value rather than in biodiversity conservation, but that does not make them any less important in urban areas. It can also be argued that highly developed urban green spaces can lessen the damaging effects of human use on other natural spaces in the city, allowing a more conventional conservation approach in other areas. A second challenge is dealing with the abundance of non-native species in urban ecosystems. Most conservationists are of the opinion that the leading cause of biodiversity loss is the abundance of alien species. This opinion is often based on the simple observation of more non-natives and fewer natives in degraded systems; however, this observation does not necessarily correlate with biodiversity when trait-based metrics are used to measure diversity and functionality. In many cases the non-natives used on the site are better adapted to the disturbed conditions than the original assemblages of native species, and a novel mixture of alien and native species are better adapted and provide more ecosystem services. The use of non-natives is not necessarily harmful if no negative impacts exist on other species or resources; however, a risk assessment should be used to consider potential risks such as movement into adjacent habitats (Kowarik 2011).

In addition to focusing on maintaining species diversity, conservation projects can focus on making landscapes more resilient and durable so they can accommodate environmental changes caused by climate change. Strategies include selecting plant species with wide tolerance ranges (plasticity), using a broad mix of species to increase functionally redundant species, and selecting species from a wide range of environments to increase the likelihood of containing particularly resilient species for the local context. Diversifying ecosystem functions will help maintain ecosystem functions for human benefits. The challenge is to realign systems so they can respond and adapt to both present and anticipated future conditions (Dooling 2015, 100–101). Recent studies are documenting examples of many more species adapting on different levels to urban life. Some are changing behaviors, others are adapting morphologically (changing shape), and others are making genetic changes or all three adaptations. All of these adaptations are the result of rapid and often extreme changes in urban areas. The good news is that as species colonize and adapt, they begin to create novel urban ecosystems that are more complex and support a greater number of species, which can improve ecosystem services by

FIGURE 10.4. Portland, Oregon: a novel ecosystem created by a green infrastructure stormwater park.

increasing biodiversity and complexity. As these ecosystems create healthy urban habitats, they can shift the urban ecological paradigm from bad to good and promote more ecological design in urban areas.

Ecologists collaborating with designers, architects, and planners is a new trend for future action and engagement in cities when designing urban habitats with a focus on evolution. In the past, and to this day, many urban spaces are designed primarily for aesthetics, recreation, and ease of maintenance, with some functionality mostly for human comfort. The urban wilderness movement, green infrastructure (fig. 10.4), green architecture, even vacant lots, are all opportunities for novel ecosystems (Kimbrough 2016). It is important to remember, however, that while all areas of a city deserve green spaces, environmental justice groups are cautioning that some ecosystem projects that focus on greening-up disadvantaged neighborhoods are leading directly to ecological gentrification with displacement due to increased rents, and an influx of wealthier households. "Just green enough" is a new strategy that focuses on environmental goals for neighborhoods that keep people in their neighborhood by making it healthier and more attractive without making the neighborhood more expensive through green gentrification. The primary goal is to scatter small parks for easier access rather than large-scale projects (Tuhus-Dubrow 2014).

Citizen Science Program: Expanding the Reach of Environmental Research with Citizen Science

The Ecological Society of America (ESA) has developed a guide for program managers to help decide if citizen science is right for their research projects and how best to design citizen science projects. The guide, titled "Issues in Ecology 19: Investing in Citizen Science Can Improve Natural Resource Management and Environmental Protection," is included as a resource in the Federal Citizen and Crowdsourcing Toolkit. The ESA released the report along with a memorandum from the White House mandating that all federal agencies include citizen science capacities (Mize 2015).

Case Study: The Environmental History of a Novel Ecosystem, Lake Claremont, Western Australia

Lake Claremont is located within the urban boundaries of Perth, a large city in Western Australia founded by British colonists. In the early 1900s, high rainfall permanently flooded the surrounding wetland, killing large trees and providing breeding ground for mosquitoes. Mosquito fish were released but proved to be an ecological disaster, feeding on the eggs of native frogs and fish. However, the decline of fish was blamed on cormorants, and shooting programs were run to kill the birds. As the leaching and run-off continued, toxic algal blooms also resulted in mass bird deaths. In 1954 the lake was filled for recreational playing fields and a drive-in cinema. In the 1960s the infill was converted to a golf course, where irrigation and fertilizer caused more algal blooms, killing more birds. In 1983, recognizing the importance of the ecological services provided by the remaining wetlands, all levels of government collaborated to develop a management plan for the lake. A Lake Claremont Advisory Committee (LCAC) was formed and a community-instigated research report was used to begin the restoration effort. Friends of Lake Claremont (FLOC) volunteers coordinated, funded, and performed infill planting to establish indigenous riparian and woodland vegetation on 10 hectares. After 10 years of restoration, the golf course is now a 12-hectare mixed-use recreational green space managed for weed control, with less fertilizer and reduced irrigation. Prior to European colonization in 1829, it is estimated that the lake region supported more than 200 bird species. Today, skilled bird-watchers from FOLC perform quarterly surveys, showing that the number of bird species has doubled from a low of 54 species in the 1960s to 96 species today (Simpson and Newsome 2017).

11

Green Infrastructure

Green infrastructure can be defined in many ways; the definitions usually hinge on scale. All natural elements inserted into a site, a landscape, or a region with the purpose of mitigating the effects of urbanization and development can be considered green infrastructure. From bioswales to greenways to wetland restoration areas, these elements help mitigate the impacts of development on the natural processes and cycles in place. Landscape elements dubbed green infrastructure either mimic natural systems or make use of natural systems to perform the same functions of "gray infrastructure" in developed areas. The latter refers to built systems such as culverts, pipes, cisterns, and other human-made structures created to replace natural systems, such as directing, diverting, and storing stormwater.

One of the natural elements most impacted by urban development is water. Its movement within sites, landscapes, and regions, and its quality and reach of distribution can be completely altered in urban areas. To mitigate the impact of urbanization on hydrological processes, cities often impose stormwater treatment measures through codes. Some regulations are legislated at the national level; for example, amendments to the United States Clean Water Act have required all states to have stormwater management plans since 1987. In addition to regulations, some states encourage best practices; for example, the Water Management Districts in the state of Florida provide stormwater management manuals that specify the use of low-impact development technologies to mitigate the effects of gray infrastructure on the environment. The use of low-impact development (LID) technologies has increased in recent years, particularly as a means to reduce stormwater runoff resulting from an increase in impervious surfaces in urban areas, which in turn increases non-point source pollution (fig. 11.1).

In addition to the advantages that green infrastructure brings to natural environments, there are social and economic benefits that result from its

FIGURE 11.1. Seattle, Washington: green infrastructure to increase pervious area and reduce the impact of development.

adoption and implementation, such as increased opportunities for physical activity in urban parks and hiking trails (Handy et al. 2002; Ward Thompson 2011), and reduced costs of drainage and stormwater systems (Braga, Porto, and Silva 2006; Scott et al. 2014; Postel 2017). Some green infrastructure offers added value because, in addition to performing ecological functions, it also provides ecosystem services. Green infrastructure can (1) connect ecosystems linking urban green areas to regional woods and forests, (2) control flooding to establish pervious green corridors, (3) mitigate contamination from impervious surfaces runoff, (4) enhance the environment-restoring riparian areas, (5) conserve wetlands and assist in filtration within the watershed, and (6) improve soil stability to control erosion.

What: Green Infrastructure at Different Scales

Cities the world over are embarking on projects to decrease their ecological footprint by restoring and maintaining the biological diversity of their ecosystems, protecting water resources, and weaving natural spaces through their urban fabric, among other strategies. The scale of projects and initiatives varies, but the final purpose is the same: allow urban development to occur in a manner that is not detrimental to the environment and take advantage of the ecosystem services offered by natural amenities.

At the site scale, the implementation of LID strategies has been successful (Clausen, Hood, and Warner 2006; Dietz and Clausen 2008; Scott et al. 2014). The reduction of impervious surfaces, retention of rainwater into rain gardens or rain barrels, and the use of bioswales for stormwater drainage offer affordable solutions. Rain gardens are an in-situ alternative to centralized retention basins for the collection and treatment of stormwater, which still depend on networks of urban infrastructure. Perception and image are significant at this scale because naturalized areas are sometimes mistaken for areas with low or no maintenance. For this reason, some cities still operate under the limitations of building codes that present obstacles to the implementation of LID strategies.

At the landscape scale, green infrastructure can provide the essential linkages between sites and regional systems. Urban parks and other green open spaces can be important nodes in a network of greenways; riparian corridors can connect different patches of natural areas; hedgerows can link forest patches separating suburban tracts of land (Forman 1995). The Patch / Corridor / Matrix model described by Forman and Godron (1986) is a useful method of analysis of landscape elements at the landscape scale. All three can vary in size, form, and texture and allow us to compare different patterns in the landscape. Forman (1995) categorizes landscapes for planning purposes in three groups: rural and agricultural land, natural resource areas, and corridors and greenways. Each one of these groups would accommodate a series of elements that could be incorporated into the landscape as green infrastructure.

At the regional scale, habitat restoration initiatives have rehabilitated landscapes. For example, beaver and associated waterfowl habitat restoration projects have rehabilitated riparian areas in regions of the United States (McKinstry, Caffrey, and Anderson 2001; Law et al. 2017). Another example of regional green infrastructure is wetland restoration. For example, several projects have attempted to protect water quality in the integrated watersheds that supply water to the 18 million inhabitants of the city of São Paulo, Brazil (Braga, Porto, and Silva 2006). One of these projects, in the Guarapiranga watershed, focused on restoring and protecting the remaining wetlands in the watershed instead of using hydraulic engineering solutions to filter water contaminated by urban and industrial uses (Frischenbruder and Pellegrino 2006). The nonstructural solution was a natural park covering an area of 24 km^2 spanning three municipalities. Other examples at the regional scale focus on connecting forest patches and remnants of naturally vegetated areas to create a continuous stretch of natural habitat that makes biodiversity possible.

So What: Lowering the Impact of Development Infrastructure

The hydrology of a site, landscape, or region and corresponding hydrologic response to disturbances represent the highest impact of development. For this reason, they are the focus of most LID strategies, particularly those targeting non-point source pollutants (Clausen, Hood, and Warner 2006; Dietz and Clausen 2008). Most infrastructure is public and modifies the natural environment. Because of the impact of infrastructure in the landscape at different scales, some scholars argue that it should be considered within a larger social-ecological system (Alberti and Marzluff 2004; Anderies 2014). Thus, integrating engineered and technological solutions is necessary for the success of green infrastructure initiatives. Low-impact development is one of these solutions; it aims to reduce the demand for water in landscapes, filter the pollution in runoff, and retain water—not only to mitigate the effects of flash floods but also to return rainwater to aquifers instead of channeling it into stormwater infrastructure.

Water and waste management strategies are widely used to lower the impact of development infrastructure, which can be achieved by retaining runoff, storing stormwater, and allowing it to evaporate and infiltrate. Both urban and agricultural runoff affect the quality of the water going into our rivers, lakes, estuaries, and other water bodies. Suburban sprawl has increased the impact by extending urbanized areas into natural reserves and agricultural lands, thus the need to protect the quality of not only surface but also groundwater so the integrity of ecosystems can be preserved. To this end, LID strategies can be implemented, including, for example, rain gardens, rainwater harvesting, green roofs (discussed in chapter 12), and permeable pavements. Stormwater runoff volume has been shown to increase exponentially as imperviousness increases (Dietz and Clausen 2008). A low-impact approach to site and neighborhood development can greatly reduce runoff by increasing infiltration, particularly in traditionally impervious surfaces such as roadways, sidewalks, and parking lots.

The city of Malmö, Sweden, became known across the globe because of its green infrastructure initiatives. Located on former agricultural flat land, most of the city had drainage problems. By creating a stormwater system that incorporates green infrastructure—for example, vegetated stormwater ponds and permeable concrete blocks in paved areas—the city was able to deal with the cycle of floods and erosion. They have had an environmental building program since 2009 that is currently used for all new developments. Malmö has consistently and successfully incorporated planning tools into all planning systems (Kruuse 2011). These green tools, such as the Green Space Factor and the Green Points System, ensure that all developments benefit from green infrastructure.

In the United States, it took 35 years from the enactment of the Clean Water Act in 1972 to the signing by the U.S. EPA of a Statement of Intent to use green infrastructure in 2007 (EPA 2015). Other signatories of the Green Infrastructure Statement of Intent include the National Association of Clean Water Agencies (NACWA), the Natural Resources Defense Council (NRDC), the Low Impact Development Center (LID Center), and the Association of State and Interstate Water Pollution Control Administrators (ASIWPCA). The intention stated in the document is that these organizations will collaborate to "promote the benefits of using green infrastructure in protecting drinking water supplies and public health, mitigating overflows from combined and separate sewers and reducing stormwater pollution" (EPA 2015, 1). Unfortunately, the regulatory structure of the Clean Water Act has hindered the ability of these organizations to use green infrastructure to its full potential (NACWA 2009). Nonetheless, considerable progress has been made in the last 10 years or so. The Low Impact Development Center, for example, has participated in several projects in collaboration with other organizations (LID Center 2019). Gradually, green infrastructure is being successfully incorporated into the built environment in many countries around the world.

Now What: An Integrative Approach

Green infrastructure has greatly improved the quality of built environments in recent years. Benefits can be compounded when different landscape elements are integrated into the system in a way that allows them to work together synergistically. Especially in the case of non-point source pollution, green infrastructure can provide the most effective way to collect, retain, and filter runoff. In parklands, for example, various types of green infrastructure are used together to achieve desired outcomes. Rain gardens and swales, restored streams, constructed wetlands, retention ponds, detention basins, permeable paving, and green roofs have been built in several public parks in American cities. In a survey of 48 selected cities, the Trust for Public Land found retention ponds in 50% of them and

FIGURE 11.2. Seattle, Washington: bioswales and rain gardens to collect, retain, and filter runoff, diverting stormwater from infrastructure.

rain gardens and swales in 71% (fig.11.2) (TPL 2016). Other types of green infrastructure were present in every city, albeit in lower percentages. Of the seven types of green infrastructure listed above, 25% of cities had four or more, and 50% had at least three types (TPL 2016). The integration of these different types of green infrastructure increases their performance and the benefits to the environment.

Another approach to balancing the impact of development is SMART Parks, where technology is used not only for provision of automated services, but also for data collection about users (Loukaitou-Sideris 2018). SMART parks may include smart irrigation systems that do not waste valuable water resources, exercise machines that produce energy used to operate the park, smart lighting that lights up dark paths and improves safety and security, and robotic lawnmowers. Some parks are defined as "water-smart parks"; green infrastructure is used to absorb runoff and hold flood waters (TPL 2016). Several cities in the United States today have parklands designed specifically to be flooded. In Austin, Texas, 11% of parklands are designed to be flooded, while in Denver 25% are. In Birmingham, Alabama, Railroad Park was designed to reduce flooding that had impaired an old industrial site, formerly a marsh, for more than 100 years (TPL 2016). The Green Infrastructure Collaborative, an alliance of federal agencies, nongovernmental organizations, and private businesses, has supported communities building water-smart parks and implementing other green infrastructure strategies (EPA 2020).

In addition to public parklands, private developments are adopting low-impact development strategies and implementing green infrastructure to decrease their environmental impact. A study conducted at the watershed scale compared two subdivisions in Waterford, Connecticut—one developed using traditional techniques and the other using LID strategies—found that both stormwater runoff and pollutant export were drastically different between the two types of developments (Dietz and Clausen 2008). The LID subdivision was found to have negligible impact on the watershed; levels of runoff and pollution in the LID subdivision were found to be the same as those in the predevelopment stage, while significant increases in stormwater runoff and pollutants were found in the traditional development.

Whether in parks or subdivisions or smaller areas within cities, green infrastructure is proven to decrease the impact of development on the environment. Introducing several landscape elements and LID strategies and combining and integrating them into urban environments allow development to occur while conserving natural features and amenities. In addition, maintenance costs are reduced, vegetation is introduced in areas usually paved in traditional developments, water quality is maintained, and aquifers are recharged. Research is continually providing evidence that it is possible to integrate nature into the city; development and conservation do not need to be mutually exclusive, and citizen scientists have a significant role to play. Having communities involved in data collection and the monitoring of performance and impact of green infrastructure can not only confirm the benefits of adopting it, but also encourage communities (and developers) to seek its implementation as a preferred alternative.

Citizen Science Program: Domestic Gardens

The city of Manchester, UK, partnered with Manchester Metropolitan University and the University of Leicester on a citizen science project in 2016. Several organizations sponsored the project and the Natural Environment Research Council provided funding. The aim was to understand how domestic gardens benefit people and nature. More specifically, the project's goal was to quantify green space across the city. More than 1,000 residents from 32 wards responded to an online survey; a prize draw to win £500 in shopping vouchers was offered as an enticement. The information collected through the survey over a six-month period was validated by remote sensing analysis (Baker, Smith, and Cavan 2018). The intention was to make better planning and investment decisions in local neighborhoods, since domestic gardens greatly contribute to urban green infrastructure.

The survey gave respondents the option to choose from ten different land covers, including buildings, impervious hard surfaces, pervious hard surfaces, bare soil, trees, shrubs, mown grass, rough grass, cultivated areas, and water.

The accuracy of garden surface estimations was higher than 75%, which allowed the researchers to extrapolate to the entire urban area and conclude that, on average, half of each lot comprised green infrastructure. The most common land covers were impervious surfaces and mown grass, occupying about 25% of a garden's area. The project concluded that around 20% of the city's green infrastructure is contained in domestic gardens and thus maintained by city residents.

This project gives communities evidence that every household counts and that residents can collectively make improvements in their neighborhoods with no hardship. It also demonstrates that local governments can develop strategies to encourage residents to increase the extent of impervious surfaces in their properties, thus contributing to the overall health of an urban area.

Expert Insight: From Flood to Urban Beach

Curitiba is a Brazilian city that became internationally known for its environmental planning initiatives. The city is the seventh largest in Brazil and during the 1990s and 2000s was lauded for offering an exemplary quality of life to its residents. The implementation of several plans and projects that conserved urban forests, riparian areas, and steep slopes are among its many accolades. Curitiba was unique in that, even while suffering from the usual woes to which urban areas in developing countries are subject, it managed to implement creative ideas at low cost to the local government and its citizens.

Nicolau Klüppel, an engineer in Curitiba, worked in stormwater and drainage projects early in his career. Later on, working in an interdisciplinary environment in the city's planning agency led him to devise novel

FIGURE 11.3. Curitiba, Brazil: most urban parks were conceived as stormwater retention areas, decreasing flood events and reducing the need for stormwater infrastructure.

alternatives to urban problems. In his own words: "I was a narrow-minded engineer, trained to build infrastructure to control floods. One day, in a conversation with my colleagues, one of them said: 'What Curitiba needs is water, but we can't have a beach at 1,000 m altitude.' I was always thinking about drainage, so I said: 'We can't have a beach, but we can have lakes that can double as retention ponds and collect runoff from increasingly large impervious surfaces.' And so we decided to have a lake in every city park." Although Curitiba does not have plentiful rivers, planners and engineers identified flood-prone areas available for conservation and built weirs to create the retention ponds and lakes (fig. 11.3). In addition to solving an environmental problem, they created a network of urban parks that are also a valuable social asset for the city and provide the population with places to exercise, practice sports, play, rest, and congregate (Macedo 2013).

12

Urban Green Spaces

There is no question that urban green spaces are important for cities. They contribute to the environmental and economic health of cities and human health and well-being. There is also no doubt about the value of ecosystems services from urban green spaces; their worth has been proven in multiple ways. The question for cities today is how, and where, to increase plants and natural areas in the urban environment. Creating additional green spaces in urban areas presents several challenges, including availability of land, development costs, maintenance budgets, and stakeholder conflicts and equity issues, but there are also some very creative solutions being implemented in many cities that can inspire others. This chapter explores strategies cities are using to insert public green spaces in their urban fabric, and some of the approaches they are using to make it happen. The chapter also presents some of the new and innovative ways in which citizens are creating private green spaces.

One thing all successful green space projects have in common is a well-defined vision and purpose for creating them, and an action plan backed by data to make a case for support and funding. The vision should articulate the type of green space desired, which can be challenging. Ian McHarg, an influential landscape architect who championed designing with nature in the 1960s, once said a green mark on a planner's map is the only thing urban green spaces have in common, confirming the great diversity of green spaces possible in a city. Although urban green spaces vary widely in size, character, function, and ecology, they can be generically described as public or private open spaces in urban areas, covered primarily in vegetation, that contribute to the well-being of humans and animals and provide ecosystem services for the broader urban area.

Green spaces can be categorized by size, location, and natural features; these are characteristics that are closely related to function. For example, recreation fields will need a specific size and shape based on the sports to be played, while nature parks and wildlife corridors can be a variety of shapes but may need to be large in size to maximize wildlife movement. Road rights-of-way will have to meet engineering requirements for safety, which will vary depending on the speed of traffic, and wetland parks will need to be sized for the volume of stormwater to be stored while the shape can vary. The features of the green space will also depend on the function; for example, dog parks need different facilities than a park created to preserve a heritage rose garden or a unique natural feature or ecosystem habitat. The users and the purpose of the urban green space also need to be taken into consideration. A focus on users has created a new trend in specialty green spaces to meet diverse needs, but the trend is not driven by any city plans or initiatives—the creative and entrepreneurial spirit of ordinary citizens usually drives trends and new ideas.

Some green spaces are created by private owners or businesses whose facilities are nature-based or require green space. Golf courses, archery ranges, cemeteries, and business parks are examples of private urban green spaces. Other examples include agriculture-based businesses such as wineries, boutique farms, and commercial plant nurseries that sometimes double as parks with greenhouses and walking trails. Restaurants are growing their own herbs and vegetables in rooftop gardens (fig. 12.1), and micro-breweries grow their own hops on small city plots. Hospitals

FIGURE 12.1. Paris, France: the green roof on a floating restaurant barge is home to several ducks.

FIGURE 12.2. Chicago, Illinois: garden at a children's hospital includes famous storybook characters and lush planting to attract birds and butterflies.

have horticulture therapy and health gardens for their patients (fig. 12.2), and homeowners are turning their backyards into their dream gardens for entertaining, relaxing, and growing food.

Public-Private Partnerships (PPPs) are also making it possible to increase the amount of green space in cities while creating innovative spaces such as zoos embedded in botanical gardens, disc golf courses, and dog parks doubling as neighborhood commons (Matisoff and Noonan 2012). Riparian parks along riverbanks, rail-to-trail bicycle corridors, and biodiversity corridors are being used for recreation areas. The broad diversity of all these innovative uses for green spaces requires the expertise of landscape architects, urban planners, sports and playfield experts, horti-culturists, land managers, arboriculture experts, nursery managers, cu-linary experts, aquaponics and aquaculture farmers, viticulturists, and horticultural therapists. Citizen scientists can also get involved; urban green spaces lend themselves to a variety of citizen science projects such as monitoring of wildlife populations in Cleveland and urban forests in Seattle and San Francisco.

What: Embedding Cities in Parks and Bioreserves

While ordinary citizens are transforming small green spaces in cities, ur-ban ecologists, landscape architects, city planners, and city officials are considering a different approach to nature in cities, one that flips the con-cept of nature-in-cities to cities-in-nature. Seeing nature in cities from this new perspective allows ecology to guide development and city form, with nature being the defining element of cities. Some ideas generated from this novel concept include urban biosphere reserves, national park cities, and reconciliation ecology.

The Urban Biosphere Reserve (UBR) is the urban equivalent of the bio-sphere reserve concept by the Man and the Biosphere (MAB) Program of UNESCO (2017) and is the first internationally recognized mission to consider cities as ecological systems. Its primary principle is to encourage a regional view of sustainable development and to acknowledge that urban planning and bioregional management are ecologically linked. UBRs are intended to contribute to the conservation of landscapes and ecosystems and to help foster economic and cultural development. Reserves also serve as research labs and support demonstration projects for conservation and development. Possible categories of UBRs include urban green belts, where the city is surrounded by the reserve, limiting further urbanization and sprawl; cluster reserves, where multiple parks and urban green spaces inside and outside the city make up the reserve; urban green corridors within the city that connect to green areas outside the city; and urban region biosphere reserves, where an entire region with small villages and towns are zoned into the reserve (Matysek 2004; UNESCO Digital Library 2004).

National Park Cities is an initiative launched by the National Park City Foundation (NPCF) in partnership with World Urban Parks and Sal-zburg Global Seminar. Its basic concept—to preserve and protect more green spaces in cities—originated in London, the first city to sign the

International Charter in 2019. The goal is to change how people think about green spaces in urban areas. National Park City is defined as "a large urban area that is managed and semi-protected through both formal and informal means to enhance the natural capital of its living landscape" (Goode 2015). A defining feature is the widespread and significant commitment of residents, visitors, and decision makers to allow natural processes to provide a foundation for a better quality of life for wildlife and people. London is building on its park heritage with more than 1,400 designated Sites of Importance for Nature Conservation, 142 designated Local Nature Reserves, and about 50 specially protected areas of national significance that amount to about 20% of land area in Greater London. An additional 20% comprises private gardens and 3,000 parks; the goal is to make more than half of London green (Goode 2015; Ledsom 2019).

Reconciliation ecology is an idea developed by Dr. Mike Rosenzweig, professor of Ecology and Evolutionary Biology at the University of Arizona (Graham 2004; Schwartz 2007). It is described as the science of inventing, establishing, and maintaining new habitats to conserve species diversity in places where people live, work, and play. The concept is to manage for biodiversity without decreasing human benefits by identifying species that can coexist with humans and creating habitats where they can survive and thrive, thus encouraging biodiversity in human-dominated ecosystems. The key is to work in cooperation with and among neighbors, neighborhoods, parks, or cities to connect small spaces and create a contiguous network large enough to support a species. One example would be to create butterfly habitat in a park or on the periphery of a golf course and encourage neighbors that abut the park to also create butterfly habitat in their yards. Several yards connected in succession become part of a continuous network to reach another park or green space (Franklin 2007).

So What: Generating Support, Partnerships, and Funding

It may be challenging to find support to increase the amount of green spaces in urban areas. Several strategies are used to get buy-in, create partnerships, and generate funding for projects, including strategically positioning the project by linking to environmental issues, looking for philanthropic sources, maximizing use and purposes for multiple partnerships, tapping into other organizations, and creating opportunities for involvement by other stakeholders. One of the most important strategies is to frame the idea or concept for a project with the aim of increasing its significance and acceptance. Positioning a project as a health, safety, and welfare concern, for example, will increase funding and partnership opportunities. Making a project part of a resilience city plan and showing how it can help mitigate climate change stressors such as flooding, sea level rise, or wildfires can greatly increase the significance of its impact. In addition, presenting evidence for the need for green spaces and linking a project to their social benefits can emphasize their day-to-day benefits (Eldridge, Burrowes, and Spauster 2019). Furthermore, appealing to philanthropic sources by choosing causes that people care about and that align with creating green spaces can also prove to be an effective strategy. For example, a proposed wildlife bridge in Los Angeles, California, was successful in generating millions in private donations by appealing to people's recognition of the need to protect wildlife.

Another effective way to garner support is to present projects as having multipurpose, multicultural, and citywide impact. Showing the potential of projects to mitigate environmental and social problems while serving as an urban green space can increase interest. For example, the High Line, an aesthetically appealing linear park built on an old elevated rail line in New York City, not only became a successful urban amenity and tourist attraction, but also solved the problem of what to do with an obsolete transportation structure while preserving history (High Line 2020). The High Line is also a good example of successful partnerships that engage stakeholders with different perspectives and tap into the experience of experts and voluntary labor. These can include private citizens; companies and foundations; local, state, national, and global organizations; government agencies and nongovernmental organizations (NGOs); not-for-profit programs; and universities, schools, and educational programs.

Case studies should be part of a green space strategy to generate support, engage in partnerships, and attract funding. Using examples of other successful projects to show potential outcomes is one of the most important components of successful marketing campaigns. One way to engage more citizens is to create large- and small-scale opportunities so everyone can get involved. Finally, a comprehensive plan is the best way to create a framework for development and conservation of green spaces. Such plans should articulate a vision and include a description of socioenvironmental benefits, along with short- and long-term schedules and budget plans (National Recreation and Park Association 2019).

Now What: New Ideas and Strategies

Recently, several novel ideas have been devised and implemented to not only increase access to green spaces in cities, but also expose the population to nature on a more regular basis. Temporary mobile green spaces, such as freight farms and large plant filters called "city trees" are literally on wheels, able to roll to different areas of the city. City parklets are also temporary green spaces that can be moved and reconstructed to be enjoyed in different areas of the city.

Freight Farms are shipping containers for growing produce that hold great promise for getting fresh legumes and vegetables to underserved neighborhoods. The containers house an indoor advanced hydroponic farm where plants grow on vertical walls in the containers and get their light energy from LEDs. The system operates independently from land, climate, or season to provide a local food source. These movable farms can support more than 13,000 plants for a steady supply, with a high production cultivation area for 8,800 plants and a seedling starter site that uses about 5 gallons of water a day (Freight Farms 2020).

The "city tree" is not actually a tree, but it was designed to function as a tree with certain ecosystem services. The large movable panel that represents a tree is embedded with moss cultures that can filter pollutants such as particulate matter and nitrogen oxides to act as an air purifier. Different mosses are connected to a fully automated water and nutrient system on a panel that removes pollution as air flows through. A ventilation fan can intensify airflow to increase the filtering effect, and the integrated sensors of the technology infrastructure collect real-time data that can be viewed through a smart phone app. The city tree is interesting because it shows the possibility of mimicking ecosystem functions in an entirely different form and size as a human-made landscape (Greencity Solutions n.d.).

Tiny parklets in San Francisco and other cities are providing dots of green throughout the city (fig. 12.3). Most are temporary pop-up parks with wood planters built in parallel parking spaces along city streets. The planters are filled with a variety of plants and usually include an attached bench for sitting. Some parks are permanently attached to the flat bed of a truck and can be moved by simply driving away. The parklet was invented in 2005 by Rebar, a team of people who design guerrilla art in public spaces. The concept inspired a global movement called Park(ing) Day with people around the world reclaiming their streets. In 2009, the San Francisco Planning Department formalized the concept by creating

FIGURE 12.3. Marmaris, Turkey: parklets along the seaside promenade provide small green spaces that add to the aesthetics and provide places to relax along the waterfront.

an experimental city program called Pavements to Parks. Through this program people get a parklet permit that allows them to convert parking spaces into mini-parks without feeding the parking meter. While only residents and building owners with the resources to build and maintain a park can receive a permit, the public is still free to use the space (Groundplay SF n.d.).

Some cities are also using derelict industrial districts and abandoned transportation corridors for new linear parks and green spaces, which is not a new concept, but the visions for spaces have become more innovative. One of the first parks to replace derelict sites was Gas Works in Seattle, built on the former site of the Seattle Gas Light Company in the 1970s. Also in Seattle, Freeway Park was built in 1976 to celebrate the U.S. Bicentennial. The Freeway Park does not replace a derelict site, but it covers a section of interstate highway that divided the city and connects the Convention Center to adjacent neighborhoods. The transformation of abandoned transportation corridors has taken on a new level of sophistication with the High Line Park in New York City and the Skygarden in

Seoul, South Korea. A common theme is the use of native plants on structures that present challenging growing conditions.

The High Line Park is a now a well-known example of re-imagining a derelict industrial site as an urban green space. The elevated rail line was opened in 1934 to bring produce and manufactured goods into the city by rail. The rail line became obsolete in the 1980s, and the Friends of the High Line, with the help of the Rails-to-Trails Conservancy, began a campaign in the 1990s to convert the former rail line into a usable park. Fundraising and rallying political and public support to overcome opposition was the biggest challenge; however, since its opening in 2009, it has become one of New York City's major destinations for tourists and residents and has generated an estimated US$2 billion in new development in the surrounding neighborhood (High Line 2020).

Skygarden in Seoul, South Korea (also known as Seoullo 7017), transformed a former inner-city highway into a "floating" garden with more than 228 species of indigenous plants, including trees, shrubs, and flowering plants, for a total of 24,000 plants. The plants grow in large concrete planters that create a variety of pathways for pedestrians and act as an educational facility; containers organized by groups of plant families follow the Korean alphabet along the pathway. The garden acts as a pedestrian connection from Seoul's main train station to surrounding city buildings and includes shops, galleries, and other amenities. At night, the containers are lit with blue light to increase plant health and growth (MVRDV n.d.).

Citizen Science Project: Canberra Nature Map

Canberra Nature Map is a repository for members of the public, volunteers, and park maintenance groups to report sightings of rare and endangered plant species in Canberra's nature parks. Participants use their smartphones or GPS-enabled cameras to photograph the unusual or rare species and report the sightings (SciStarter 2019b).

Case Study: Gardens by the Bay, Singapore

Gardens by the Bay is a 250-acre park of reclaimed land in the Central Region of Singapore. The garden is part of Singapore's national plan to transform the Garden City into a "City in a Garden." The goal is to raise the quality of life in Singapore with more plant material. The garden has become a national icon with three waterfront gardens and the largest glass greenhouse without columns in the world. The most iconic feature is the Supertree Grove, where the metal tree structures are designed to vent hot air from the conservatories and circulate cool water back. The trees are covered in ferns, vines, bromeliads, and orchids and are fitted with technology to mimic the ecological functions of trees. Photovoltaic cells harness solar energy, and the shapes of the trees collect rainwater. The garden also has an "edutainment" goal to help visitors learn about plants with themed gardens such as the Heritage Garden and Colonial Garden (Gardens by the Bay n.d.).

13

Urban Structures

Urban structures comprise buildings and hardscapes in the urban environment. With increased attention being paid to sustainability in cities, urban structures have evolved, and great progress has been made in the last 20 years or so. Green buildings and policies mandating or encouraging them have proliferated in our cities. New materials and technologies have been developed to decrease the amount of debris used in construction and of energy consumed, not only in the production of construction materials, but also in the life of the building. Green roofs and green walls have become common features of green buildings. Green roofs are efficient at reducing water runoff, absorbing and filtering rainwater, and diverting it from stormwater systems. Green walls are not only decorative, but also assist in retaining water and purifying air quality, especially in indoor environments (fig. 13.1).

Some sustainable construction materials are reused or recycled and, in the same way they can make buildings more sustainable, new hardscape materials can make landscapes more sustainable. For example, permeable paths and pavers have less impact than impervious asphalt and concrete surfaces; by allowing water to percolate and be absorbed back into the ground, they reduce runoff while helping to restore aquifers. Percolation also decreases the level of contamination in stormwater runoff as the water filters through the soil before reaching the water table. In places where stormwater infrastructure cannot endure flash floods and extreme weather events, green roofs, green walls, and sustainable hardscape materials can help mitigate potential disasters.

What: What Is a Sustainable Built Environment?

Several elements at different scales work in concert to create a sustainable built environment. Buildings, comprising the bulk of a built environment,

FIGURE 13.1. New York City: green wall in the Lincoln Center for the Performing Arts.

have the most impact. Infrastructure and other hardscapes surrounding buildings can add to or mitigate their impact. All human-made elements of a built environment present challenges to urban dwellers seeking to maintain and increase sustainability, and the selection of materials can make a significant difference. Green builders use life-cycle assessments to measure the impact of buildings, including the materials used to build them. Taking the entire life cycle of a product into consideration when analyzing the impact of any product or structure is commonly referred to as a "cradle-to-cradle" approach (McDonough and Braungart 2002). This method considers not only production and disposal of materials, but also their functional life. For example, most construction materials will have an impact on future energy use in a building, the quality of indoor air, and the costs to operate and maintain the building.

The sites on which buildings sit also impact the quality of the built environment. Green infrastructure, covered in chapter 11 of this volume, makes a valuable contribution to achieving a sustainable built environment. Both private and public structures and services make up the built environment; while buildings usually occupy private spaces, infrastructure connects private sites to public networks. Urban environments are usually

FIGURE 13.2. Seattle, Washington: Freeway Park, designed by landscape architect Lawrence Halprin in the early 1970s.

dominated by building sites where concrete and asphalt prevail (fig. 13.2). The design criteria influencing built environments can determine not only the sustainability, but also the resilience of these complex systems (Anderies 2014). It is necessary to balance specific performance goals, such as those defined by life-cycle assessments, with the impact of these goals on the built environment.

A sustainable built environment should contain structures whose impact is minimized by the use of new technologies and techniques, new materials, reuse and recycling of existing materials, and a comprehensive approach that takes the entire life cycle of structures into consideration. In addition, the infrastructure connecting buildings and forming a sustainable urban fabric should also minimize impact. Sustainable built environments do not happen by default or accident; they are designed and purposefully made in ways that reduce the impact of human-made structures on the natural environment. In Ådland, Sweden, for example, the goal of a new development is to have zero-emission buildings in the entire area named Zero Village Bergen. To achieve this goal, dwellings are being designed and placed on the site to maximize solar and daylight access; building envelopes have dynamic thermal properties and optimize solar gains (Hestnes and Eik-Nes 2017). The development also uses water treatment, heating, and mobility strategies designed with the ambition of achieving zero balance.

So What: Evolving Urban Structures

The adaptation of urban structures to the current need to restore natural environments, mitigate the effects of climate change, and improve quality of life and well-being in urban areas is paramount. Technological innovations have contributed to the evolution of urban structures with new materials being used in the construction of buildings and infrastructure. The operation, maintenance, renovation, and repurposing of buildings and urban infrastructure have also contributed to the evolution of the built environment. The first generation of urban structures seeking to create a sustainable built environment were popularly known as green buildings (Kibert 1999). Today, more advanced communities have made zero-emission buildings their goal. Zero-emission buildings not only reduce energy demand, but also balance their demand with power generated from renewable resources (Hestnes and Eik-Nes 2017). With the evolution of materials and construction techniques and in light of new knowledge about climate change and its consequences to the planet, zero-emission buildings are becoming a reality, and zero-emission built environments could be achieved as early as 2050.

In the past 20 years, green buildings have become the norm in communities seeking to achieve new levels of sustainability. Changing water and solid waste management practices that not only preserve natural resources, but also control the way in which disposal occurs in urban areas have been implemented in virtually every city across the globe. In addition to government regulations, several organizations have developed standards to guide professionals in their choices of materials and techniques applied in the built environment. Most of these organizations provide third-party voluntary certifications that verify manufacturers' and builders' claims and qualify the environmental performance of products and buildings (Kibert 1999). One such organization is the U.S. Green Building Council (USGBC), which has created the LEED certification system. Leadership in Energy and Environmental Design (LEED) is a rating system that evaluates buildings, taking into consideration their design, construction, operations, and maintenance. In addition to LEED in the United States, and in the same spirit, other countries have developed their own rating systems to encourage the development of green buildings. For example, the Building Research Establishment Environmental Assessment Method (BREEAM), implemented in the UK before LEED in the United States, was the first environmental accreditation system for buildings, and it is used in more than

50 countries. Building Environmental Performance Assessment Criteria (BEPAC) is used in Canada, and Green Star is used in Australia.

Life-cycle assessments (LCA) take a more holistic approach to built environment sustainability. LCA concepts consider the development of the site, the infrastructure required to sustain the facility on it, and the facility itself. In addition, they consider the products, processes, and operations during the active life of a building, and its eventual disposal, including refurbishment as a way to extend its life, or recovery of its reusable and recyclable components upon demolition. Through life-cycle analysis, it is possible to keep energy consumption in check, reducing the total amount of urban waste generated in built environments. Environmental labels are often used in conjunction with performance and life-cycle assessments (Kibert 1999). The nonprofit organization Green Seal, for example, provides life-cycle-based standards and certifies products that meet specific criteria to reduce their environmental impact, including reducing toxic pollution and waste and conserving resources (Green Seal 2019). They certify products that meet strict science-based standards and issue "ecolabels" for such materials as windows, thermal insulation, paints, and adhesives that improve the energy efficiency of facilities as well as the health of their users. Life-cycle-based parameters also aim to reduce or eliminate the use of materials that deplete ozone and contribute to the harmful causes of long-term climate change. In addition to third-party certifications and ecolabels, there are countless guides, handbooks, and technical reports that assist in conducting life-cycle analyses and in determining the environmental impact of buildings. The increase in number of performance criteria drove the evolution of comprehensive building environmental-assessment methods, which have allowed builders and cities to be more environmentally responsible and reduce the detrimental environmental impact of urban structures.

Now What: The Next Generation of Urban Structures

The evolution of urban structures has improved our built environments, and some characteristics, once considered innovative and unique, have become the norm. In addition to green and zero-emission buildings, green infrastructure has been integrated into several concepts being employed to deal not only with urban water cycle management, but also with the growing impact of extreme weather events related to climate change. Different strategies have been adopted in different countries: Low-Impact

FIGURE 13.3. Beijing, China: green patches everywhere contribute to the district-wide approach of sponge cities.

Development (LID) in North America, Water Sensitive Urban Design (WSUD) in Australia, Low-Impact Urban Design and Development (LIUDD) in New Zealand, Sustainable Urban Drainage Systems (SUDS) in the United Kingdom, and the Climate Proof City (CPC) in the Netherlands. Chinese cities have embraced a district-wide concept—"sponge cities"—to adapt urban structures so that built environments will contribute to the restoration of natural environments (fig. 13.3). The basic concept combines well-known principles from LID, WSUD, LIUDD, SUDS, and CPC strategies into an integrated urban water management system and applies them to entire districts, some in existing urban areas, most in new planned cities and suburbs.

From a landscape architecture perspective, the sponge city concept is based on working with nature rather than trying to control it (Saunders and Yu 2012). In sponge cities, parks and other natural areas are designed to be flooded during the rainy season and then used for other purposes during the dry season (fig. 13.4). This approach not only decreases the need for stormwater infrastructure, it also allows large areas to remain pervious,

FIGURE 13.4. Beijing, China: parks and other natural areas in sponge cities are designed to work with nature.

which helps return water to the aquifers and to create vegetated areas and habitat for wildlife. Natural areas are just one component of this comprehensive approach; urban structures such as green buildings with green roofs and walls, permeable pavements, and other structures that harvest rainwater are also part of this strategy.

Many sponge city projects were part of new cities built in China to accommodate its staggering rate of urbanization; China's urban population grew by 10% between 2005 and 2015 (Jia et al. 2017). Because China decided to use the sponge city concept at a national scale, as part of its 2007 "Ecological Civilization" platform (Dai et al. 2018), the implementation of sponge cities became a significant urbanization strategy. The Sponge Cities Plan, launched by the Chinese government in 2014, intends to transform urban structures in 657 cities by integrating green and gray infrastructure (Jia et al. 2017). The first 16 pilot cities were selected in 2015, and an additional 14 in 2016. The 30 pilot areas varied in size, from 7 square miles (18 km^2) in Xixian to 30 square miles (80 km^2) in Shanghai, and only 20 of the 30 cities have wastewater recycling included in their project goals (Li et al. 2017). A financial scheme including subsidies, matching funds by local governments, and public-private partnerships is supposed to cover the US$1.5 trillion total cost (Jia et al. 2017).

One of the 16 cities chosen for the original pilot program was Wuhan, the largest city in Hubei Province. Wuhan's 8.5 million people occupy an area of 8,500 km², about 3,300 square miles (Dai et al. 2018). Before the rampant urbanization of the 1990s and 2000s, Wuhan was dubbed "the city of hundreds of lakes"; however, only 30 of its original 127 lakes remain (Dai et al. 2018; The Guardian 2019). Although the majority of Chinese cities flood regularly, extreme weather events triggered flood-prevention governmental action; disastrous flooding occurred in the Yangtze River Basin in 1998, in Beijing in 2012, and in Wuhan, Nanjing, and Tianjin in 2016. The pilot project for Wuhan—covering almost 15 square miles (40 km²) of the city area and including 455 sponge projects costing US$2.4 billion—was accelerated because several people died in 2016 when the city was afflicted by floods (Dai et al. 2018; The Guardian 2019). In addition to the sponge city areas, Wuhan has several parks where rainwater is harvested and reused in park sprinkler systems.

Sponge cities do not rely solely on passive green infrastructures, such as bioswales and permeable pavement, which have been extensively used as low-impact development strategies in the United States and Canada. While LID mimics natural processes, the structures required to implement the sponge city concept are massive given the size and population of Chinese cities. Once absorbed through permeable pavement, constructed wetlands, rain gardens, and bioswales, water is directed through underground tunnels to storage tanks and then reused to irrigate areas in times of drought. Sponge cities became necessary in China because urban structures replaced natural areas at a fast rate; rivers and wetlands were filled and paved over to accommodate the growth of urban areas. In addition to mitigating the effects of rapid, unsustainable urbanization, cities adopting this strategy, particularly those in flatlands and coastal areas, will be in a better position to deal with sea level rise.

The large number of sponge city projects in China will provide researchers an infinite number of opportunities for life-cycle assessments. Citizen scientists could assist researchers with the collection of performance data in sponge cities. Li et al. (2019), Lin et al. (2018), and Mei et al. (2018) have attempted to quantify the performance of LID practices, carbon emissions, and cost, respectively; however, the recent implementation of projects does not yet allow measurement of their true impact. It will be important to collect data as these projects are completed so that we come to understand the efficacy of these urban structures and are able to evaluate their performance within a life-cycle assessment framework.

Citizen Science Program: Mohawk College, Ontario

Canadian educational institutions are inspiring not only their students, but also the citizens in the communities where they are located to make the built environment more sustainable. Mohawk College, in Hamilton, Ontario, instituted a Sustainability Initiatives Fund to support the more than 30 initiatives in their environmental management plan. One of these initiatives, in partnership with students, is a pollinator-friendly green roof on their library building. In addition to growing native species to attract pollinators essential to the health of food systems, this 2,400 ft² rooftop garden is engaging students and community members as citizen scientists.

This new learning environment was jointly funded by the College and the World Wildlife Fund (WWF). Mohawk was the only Canadian college to be awarded a grant from WWF's community grants. The intention of the granting agency is to strengthen citizens' connection with nature and engage them in the creation of healthy habitats, not only for pollinators, but also for wildlife in general. Students and citizens can learn about rooftop gardens, native plants, bees, butterflies, hummingbirds, and beetles— some of which are on Ontario's species-at-risk list—and then replicate the idea throughout green spaces in their communities (Mohawk 2019).

Projects: Green Roofs Pioneers

The City of Chicago in the United States and the City of Toronto in Canada were the first cities in North America to adopt green roofs as a biophilic strategy. Chicago led by example, making City Hall its first building with a green roof in 2001, and was awarded an American Society of Landscape Architects (ASLA) 2002 Professional Merit Award. Toronto's City Council adopted a Green Roof Strategy in 2006 and was awarded the Federation of Canadian Municipalities FCM CH2M Hill Sustainable Community Award in 2007.

Chicago's primary objective was to address the Urban Heat Island effect in the wake of the 1995 heat wave that caused the deaths of hundreds of citizens. The US$2.5 million project was made possible by a settlement reached by the city with the electric utility Commonwealth Edison; part of the settlement was allocated to the Urban Heat Island Initiative, managed by the city's Department of the Environment. Chicago City Hall's green roof is a 20,300 ft² (1,900 m²) native garden on a 38,800 ft² (3,600 m²) rooftop. Collected rainwater is used to irrigate the garden, but there

FIGURE 13.5. Vancouver, Canada: 6 acre (2.4 ha) living roof of the Vancouver Convention Centre, a LEED Platinum building. This green roof is the largest in Canada and the largest non-industrial living roof in all of North America. It features 350,000 plants, 40,000 bulbs, and 10 different species of native grass and herb seeds; it is mowed once a year.

is also an irrigation system to supplement natural rainfall during the dry months. Since the implementation of this first green roof, which lowered the building's utility bills by US$5,000 a year, more than 500 vegetated roofs were built in Chicago, totaling almost 6 million square feet of green roof coverage (City of Chicago 2019).

Toronto's approach was more gradual. The city adopted a Green Roof Bylaw in 2009 after commissioning a study by Ryerson University that

FIGURE 13.6. Halifax, Canada: a sedum green roof combined with a photovoltaic array.

determined green roofs would optimize energy use and have a positive impact on stormwater management. The city used communications and marketing, financial incentives (EcoRoof Incentive Program), public education (workshops and training), and a streamlined development approval process to promote and implement the strategy. In the pilot phase of the program, 32,000 ft² (3,000 m²) of green roofs were built. In 2007, an additional 14 roofs were approved. Since launching in 2009, this bylaw has enabled the construction of more than 350 cool roofs and more than 70 green roofs, saving close to 1.6 million kWh in energy annually. Not only has this produced significant energy savings, but it has also diverted upwards of 3.4 million gallons (13 million liters) of stormwater per year (City of Toronto 2019). Chicago and Toronto have pioneered the green roof movement in North America and have shown that leading by example and enabling developers to adopt sustainable practices can have a significant impact in urban environments and in the economy. Many other cities today are encouraging the implementation of green roofs not only to save energy and retain rainwater, but also to nurture an overall more sustainable built environment (figs. 13.5 and 13.6).

14

Risk Assessment and Climate Change

Climate change, growing populations, and environmental degradation have created new urgency for cities to consider risk assessment policies and climate-proofing strategies. A key concept for urban development policy is to consider how cities will be able to adapt or make adjustments to environmental pressures that can either mitigate harm or exploit opportunities. Urban planners think about adaptation on several levels: (1) anticipatory adaptation—to make adjustments before environmental impacts are detected; (2) autonomous adaptation or spontaneous adaptation—a response that happens without planning due to ecological changes in natural systems or economic changes in human systems; and (3) planned adaptation—deliberate action, such as a policy decision, to create a desired condition before conditions change or are about to change (Douglas and James 2015c, 368). Most cities engage in a variety of all three levels of adaptation, depending on the current and anticipated type and severity of environmental change and the impact on human health and well-being. The shift toward adaptation as an urban protection strategy came about with the evolving view of socioecological systems and the emphasis on nature and people in cities. Adaptation, rather than preservation or conservation, is a different way of thinking about resilient cities, especially when applied to plant communities and understanding the ecological processes in cities. This requires interdisciplinary approaches in ecology and sociology from a multitude of disciplines, including urban ecology, political science, public health, economics, and social welfare. Urban planners typically consider two different approaches to using natural processes to prepare cities for future impacts. The first approach includes building new urban areas on previously used agricultural sites, or old industrial and transportation sites. These areas provide the opportunity for appropriate design from the beginning, including adequate green space, public transportation, energy efficiency, and smart technologies. The second approach is retrofitting existing built infrastructure and green areas and changing people's behavior to reduce energy and resource consumption. This can be done by upgrading buildings with green roofs and renewable energy, renovating transportation facilities and creating more green space. The challenge with older cities is dealing with environmental legacies and entrenched human behaviors. However, this challenge also presents the opportunity to gather and use important information from the environmental history of the area to guide new ideas (Douglas and James 2015g, 400–401, 417).

Citizen scientists have been involved in climate change projects for some time. It has been estimated that data collected from citizen scientists have contributed to 77% of recent studies on birds and climate change. Examples include a current project, SmartFin, in which surfer volunteers use surfboard fins with embedded sensors to measure ocean pH and temperature while surfing. The data are being used by scientists worldwide for ocean research (Pickett 2017). Another project, RinkWatch, asks citizens to track climate change by observing their backyard skating rinks and documenting the location and number of days they can skate (Petri 2017; RinkWatch n.d.). In the United States, both the National Oceanic and Atmospheric Administration (NOAA) and the National Aeronautics and Space Administration (NASA) sponsor citizen science projects connected to climate change, including collecting community snow observations, mapping floating forests to identify trends in global kelp cover, and measuring variations in water storage in lakes (NASA 2018). NOAA has teamed up with the Cornell Lab of Ornithology for the Communicating Climate Change (C3) initiative, using online media to show impacts of climate change through citizen science (NOAA Climate n.d.).

What: Environmental Risks and Future-Proofing from Climate Change

Climate change impacts faced by cities today include (1) changes in morphology, such as rising sea levels and loss of urban forests and green spaces; (2) loss of function due to less biodiversity, extreme heat, and polluted water and air; and (3) compromised human health and well-being due to depleted water supplies and food shortages caused by loss of agricultural land. Other human-made and natural risks such as floods and earthquakes, pest and disease infestations, and chemical pollution can also impact cities (Douglas and James 2015c, 372). Green urbanism is a general

FIGURE 14.1. Salerno, Italy: concrete forms absorb waves and provide coastal protection from severe storms along the waterfront of Port Beach in Campania, Salerno.

concept embraced by urban planners to mitigate climate change effects. Several future-proofing approaches fall under the umbrella of green urbanism, including resilient cities, smart cities, and eco-cities. Characteristics of a future-proof city include (1) healthy coasts and green spaces with provisions to protect ecosystems; (2) energy-saving green buildings, sustainable transport, and efficient waste management; and (3) a secure water supply and clean energy (Atkins 2012). Implementing future-proofing requires a series of actions and cooperative planning. The first step for any city is to determine risks, vulnerabilities, and capacities in the city and identifying potential solutions. Conditions to identify include areas with greatest flooding risks, water and food security threats, and the energy trajectory of the city. It is also important to identify who is most vulnerable to environmental hazards and the policy options for responding to risks. The key is to plan for and manage uncertainties while still providing social and economic benefits (Atkins 2012).

Resilient cities use weatherproofing, waterproofing, and heat-proofing as a suite of future-proofing strategies. Weatherproofing mitigates unpredictable and severe weather, including strategies to create hurricane-proof urban forests and coastal protection (fig 14.1). Waterproofing includes flood control and preventing saltwater intrusion, and heat-proofing includes strategies to reduce the urban heat island.

Smart cities employ a suite of urban planning strategies that focus on the use of technology to climate-proof existing buildings and adapt infrastructure such as water and wastewater systems. Smart cities also reduce resource consumption by using technology to improve energy, transportation, and utility infrastructure, a key difference from resilient cities. Technology is also used to establish emergency systems and disaster risk-management programs to reduce health impacts to people. Eco-cities combine the information and communication technologies of a smart city with low carbon and resource-efficient development to mitigate threats. A key activity is expanding the green infrastructure with more parks and wetlands (Douglas and James 2015g, 411–413).

The overall goal of green urbanism is to use urban design and environmental strategies to make cities more energy and resource efficient and lessen environmental degradation. A major principle of green urbanism is designing for change with flexibility in how we use structures, infrastructure, transportation, public spaces, and green spaces in cities in the future. Creating resilient cities that can withstand change requires an understanding of environmental and human-made hazards and using risk assessment to develop policy and management plans.

So What: Urban Climate Change Impacts on the Environment

Urban climate hazards are climate-generated stressors that affect urban ecosystems. Slow climate changes include increasing temperatures, sea level rise, and changes in rainfall patterns. Abrupt changes occur with severe events such as wildfires, heat waves, storm surges, and floods (fig. 14.2). The impacts on human well-being are well documented, and efforts to future-proof cities focus primarily on the human condition. Many of the solutions and strategies being proposed include the use of natural areas and nature to mitigate habitat and biodiversity loss that help prevent flooding and extreme heat. Although these natural areas are considered protective strategies, the animals, insects, and vegetation in those areas are also impacted. Examples include the urban heat island that can alter flowering times for plants, which then affects plant diversity, rates of carbon sequestration, and productivity of plants. Animals also respond to heat by changing breeding cycles and migration patterns. New animals may appear in cities where warmer temperatures year-round reduce the need to

FIGURE 14.2. Miami, Florida: flooded parking lot at Fairchild Botanical Gardens after a typical summer storm. Flooding has become common with king tides and more severe storms.

migrate in the winter. Changing species distribution with increased heat levels have already been documented. Cities in tropical zones may lose species as plants and animals shift northward, and the introduction of new species may reduce native species. Warmer cities may also experience more invasive species and pests that were previously limited by the cold.

When unpredictable water sources and heat extremes affect urban green spaces, there is a cascading effect on many organisms. Heat and drought are particularly damaging to plants and animals. Earthworms and other invertebrates are affected by the dry/wet cycles of soils caused by drought conditions that are often followed by over-irrigation in response to droughts. Small reptiles and amphibians, such as frogs, are especially vulnerable; lack of rainfall can alter their breeding season since they use rain as an indicator of when to lay eggs. Small urban mammals often change courting and mating behaviors during breeding season in response to weather extremes (Douglas and James 2015c, 372–373). Less groundwater due to drought can also increase salinization, affecting the quality and quantity of water available to plants and animals, which ultimately affects biodiversity and productivity. In coastal areas, increased salinization and reduced groundwater recharge decrease habitat quality, especially for plants not saltwater tolerant. Flood hazards can include inundation,

FIGURE 14.3. Harbin, China: constructed stormwater pond/wetland in new development area of Harbin that supports the sponge city concept.

animal drowning, sediment movement (erosion), and disrupted soil processes. Discharge from impervious surfaces often carries large quantities of nitrates from spilled fertilizers into water bodies. The result is large algal blooms that suffocate fish and inhibit plant growth (Feinberg and Hostetler 2013).

Urban ecosystems are more vulnerable because, in addition to climatic stressors, they are also exposed to non-climate stressors from high concentrations of people. When assessing vulnerability of ecosystems and deciding on best adaptive strategies, it is more cost effective and useful to assess at the landscape level, as opposed to the single-species level. Managers should aim to minimize threats to biodiversity, reduce constraints to adaptation, and reduce exposure to climate changes. Examples of adaptive strategies could include planting species more tolerant of heat and drought and introducing more urban green spaces and constructed wetlands (fig. 14.3). Preserving agricultural land in outlying areas will also reduce heat loads and flood hazards and improve air and water quality. Although connected green corridors are often desired over fragmented green spaces, the total percentage of green space is the most important factor for resiliency and probably more practical to implement as an adaptive strategy (Rosenzweig et al. 2015, 344, 347–348).

Now What: Implementing Strategies for Resilient Cities

Water is the most pressing issue for most communities experiencing climate change impacts. Flooding, saltwater intrusion, rising sea levels, drought, and loss of water quality and scarcity are problems for nearly every city around the globe. Some urban planning strategies currently being used include sponge cities, water plazas, and blue/green wedges to improve conditions of both too much water and not enough water. The Sponge City Concept (SCC, which originated in India and Vietnam) is being used in several countries (fig. 14.4). The concept incorporates strategies from low-impact development, sustainable urban drainage systems (SUDS), and green infrastructure (see chapter 13, Urban Structures, for more discussion on sponge cities) (Zevenbergen, Fu, and Pathirana 2018, 2–4).

While the sponge city approach relies on natural processes, other cities have turned to solutions using built structures to hold and reuse rainwater. Water plazas are an innovative approach to handling excess stormwater in the urban core. The first full-scale water plaza was built in the city of Rotterdam, Netherlands in 2014. An unused urban plaza was redesigned with low areas that can retain up to 2 million liters (528,344 gallons) of water during storms. During the dry season, the flat bottoms of the shallow

FIGURE 14.4. Harbin, China: large stormwater park that integrates several techniques for infiltration and reuse of stormwater.

recessed "pools" are used for sports, such as basketball, skateboarding, and an outdoor theater space. The project was part of Rotterdam's climate-proof initiative, and the city became one of the first to join the Rockefeller Foundation 100 Resilient Cities network (The Stormwater Report 2014).

Other cities are using the concepts of blue/green wedges to take advantage of existing water bodies or to create human-made water storage systems such as wetlands. Blue/green wedges (sometimes called green fingers) are a type of urban green corridor or biodiversity corridor being used to waterproof and heat-proof cities. The name comes from the characteristic slice-of-pie shape of the green space which may also include a river (the blue wedge). Wedges are typically narrow in the city core where high density restricts the form, then widen as they move from the core to the edge of the city and into rural areas where more space allows for greater width. The wedges act as conduits into and out of the city and provide several ecosystem services related to climate change and future-proofing of cities such as supporting biodiversity, flood control, heat island cooling, and providing health and well-being services (Douglas and James 2015c, 383; The Nature of Cities 2014).

Managed retreat for cities is a strategy being considered by some cities in preparation for climate change impacts that could make cities uninhabitable. It includes the process of demolishing existing development and relocating people, which is a costly, political, and unpopular concept from which most cities would see little gain in the short term. Retreat also includes protecting undeveloped natural areas as a means to waterproof other areas (such as green spaces to absorb flood waters). Retreat strategies include:

- eliminating existing development through property acquisition, which requires willing sellers;
- mandating removal of development through legal and regulatory policy, which requires fair compensation;
- prohibiting property protection actions by owners, forcing voluntary retreat;
- restricting or prohibiting future development or post-disaster rebuilding, with the goal to improve ecosystem services to improve resiliency; and
- limiting support and services for development, such as not providing roads, water or electricity to high-risk areas (Plastrik 2018, 2, 4, 6, 11, 12).

Some managed retreat plans also include resettlement and identification of receiving communities for relocation. Currently most cities are letting market dynamics respond to the risks in ways that property abandonment occurs without management, such as insurers and financial institutions refusing to insure or finance properties or climate-proofing projects. One problem already seen with this approach in coastal communities is the gentrification of older, inland neighborhoods as wealthy homeowners sell coastal properties and force relocation from less wealthy neighborhoods by driving up real estate values.

Ongoing research in urban systems shows the beneficial and cost-effective outcomes of investing in urban ecosystems and biodiversity to improve climate adaptation. Future planning and policy development should: (1) invest in green infrastructure planning and ecosystem-based adaptation; (2) incorporate both the monetary and nonmonetary values of biodiversity and ecosystem services into the cost-benefit analysis; (3) recognize the social-ecological relationships that co-produce ecosystem services; and (4) use collaborative, cross-boundary biodiversity and ecosystem research to develop more nature-based solutions (Rosenzweig et al. 2015, 366).

Many organizations and institutions have been established worldwide to help cities cope with climate change and develop strategies for future growth. International programs include:

United Nations Educational, Scientific and Cultural Organization (UNESCO)
International Network Urban Biodiversity and Design (URBIO)
United Nations University Institute for Advanced Study of Sustainability (UNU-IAS)
International Union for Conservation of Nature (IUCN)
Local Governments for Sustainability (ICLEI)
United Nations Environment Programme (UNEP)
Secretariat of the Convention on Biological Diversity (SCBD)

Initiatives include TEEB for Cities, (The Economics of Ecosystems and Biodiversity), Landscape Connectivity and City-Region Planning Initiative, Advisory Committee on Cities and Biodiversity, and the Urban Biosphere Initiative (Urbis Initiative) (Rosenzweig et al. 2015, 363).

Citizen Science: iSeeChange

iSeeChange is a citizen science project that helps communities document weather in their own backyards. The motto is "you are the expert on your own block," and the intent is to help communities that are most vulnerable to the effects of climate change. Participants contribute descriptions of weather and climate, including photos combined with weather and satellite data to discover climate change patterns and make recommendations about resiliency strategies that will most likely allow a community to withstand climate change stressors. The project is partnering with NASA and NOAA to correlate community experiences with space-based observations (Bryan 2017).

Projects: New York City's Staten Island Bluebelt

The Staten Island Bluebelt stormwater management system is considered one of the best examples of an integrated ecosystem-based adaptation and disaster risk reduction response system. The Bluebelt, a system of created wetlands that have been developed since the 1990s, is a model for providing ecosystem services such as improved water quality, wildlife habitat, and stormwater management. The primary goal of the Bluebelt project was to preserve the remaining wetlands on Staten Island and to provide stormwater infrastructure. Since 1995 more than 50 sites have been developed in the Bluebelt at a cost savings of US$30 million compared with the cost of a conventional piped stormwater system. The facilities are designed to be a treatment train where water flows through a forebay to an extended wetland with native wetland plants and soils. Bioengineering techniques are used to stabilize slopes while native trees and plants grow. Wetland trees and shrubs are excavated from local development sites with the full soil profile intact to improve transplantation success. Habitats are created from old trees, brush piles, and boulder piles. The dams and bridges for the trail system are built from fieldstone to fit the character of historical structures. Water quality monitoring shows that the nutrient removal exceeds national standards for pollutant removal, and it is estimated that the system has saved the city more than US$80 million compared with a conventional stormwater system. Other NYC boroughs are now using the concept under the NYC sustainability plan, "PlaNYC 2030" (Rosenzweig et al. 2015, 328–330).

PART III

ECOLOGY *FOR* CITIES

Humans and Nature

Chapters in Part III describe the human relationship with ecology in cities and the role humans play in the protection and creation of ecological systems that are critical for cities to maintain ecosystem services and sustainability in the future. The first three chapters collectively describe how humans have manipulated, managed, and used our natural resources historically in urban spaces and the modern-day approach to managing our urban resources. Several historic phenomena influenced the look and use of green spaces in cities today, including the concept of collecting plants from distant places, the invention of irrigation, and the influence of artists who painted nature scenes. Humans' early version of nature in gardens is often depicted in ancient murals as highly organized collections of plants in private gardens for the wealthy. As cities expanded, gardens and designed green spaces became more common with the invention of irrigation, which made it possible for humans to have extensive gardens and grow edible plants, allowing them to settle permanently and expand their cities. As cities grew, human perceptions of an idealized nature changed as they sought to design green spaces for their leisure as more perfect versions of natural areas, often influenced by prominent landscape artists of the time and the emergence of professional garden designers (chapter 15). While history continues to influence modern-day approaches to urban development, the new science of urban ecosystems (chapter 16) provides a significant step forward for environmental and land use legislation. Strategies that relate to protection and management of natural resources when developing urban areas are common today, although they vary widely in intent and enforcement. Current strategies for global urban sustainability include the role of agencies and agreements, such as the Kyoto Protocol and the Paris Accord, to mitigate climate change stressors that are impacting urban areas around the globe (chapter 17).

Part III continues with chapters on humans' need for nature (biophilia), the emotional bonds that create meaning and attachment to nature, and landscape behaviors, driven by values and norms, that impact nature in urban areas. Biophilia, attachment, and landscape behaviors are discussed to show how our need for nature presents an obligation for more sustainable practices in future cities. Sustainable practices are often based on the value we put on nature and our willingness to act ethically and responsibly in our treatment of nature. Biophilia (chapter 18) is the concept that humans are inherently attracted to nature, from which they receive numerous physical and mental health benefits. In urban areas, biophilic design, both indoors and outdoors, can improve quality of life and well-being. Place-making is described in Chapter 19 as a common activity in which people engage to create a space that has personal meaning and reflects community values and norms. The emotional bond people have with nature is a significant concept for the design of urban green spaces and the actions and activities in which humans engage to create, care for, and protect nature in cities (chapter 20). People focus on different aspects when we value nature in urban areas. Cultural beliefs and ethics are presented as the moral ground for social and environmental policies for sustainable cities, and valuing ecosystem services is discussed as the approach taken by many cities in the planning and development process. Humans place a perceived or assigned value on nature through personal values, ethical behavior, moral beliefs, and cultural attitudes (chapter 21) that lead us to practices that protect or misuse nature. We may value nature simply because we love it and believe nature has rights, because we believe it's the right thing to do, or because we enjoy the benefits and provisions of nature that promote our health and well-being (chapter 22). Several valuation processes for ecosystem services, from ecological and cultural perspectives, have been developed to value nature. However, there are many challenges, both moral and ethical, with incorporating natural elements into economic systems for profit. The U.S. General Services Administration recently published a white paper that illustrates how all these "values" can be addressed through landscape performance indicators when commissioning a site (chapter 23). Decision-making depends on information that comes from landscape performance data, including social, environmental, and economic data that can support policy and a vision for community growth that protects the environment. Indicators can include metrics for water savings, soil storage capacity, vegetation protection, stormwater and temperature control, aesthetic value, and many other functions that improve human welfare (Mortice et al. 2019).

Chapters 24 and 25 describe the impact of urban nature, or lack of it, on human health. Healthy and sustainable neighborhoods are the product of policies that protect the environment and meet our mental and physical health needs, which is discussed in the context of environmental justice as a concept to ensure equitable access to ecosystem services and protection from toxic environments.

Environmental justice (chapter 24) addresses health, safety, and social equity in the urban environment based on the distribution of both nature-based amenities and environmental health hazards. Analyses show that distribution of both are tied to affluence and result in the same outcome: poor, minority neighborhoods often have fewer nature-based amenities and more hazards, and affluent neighborhoods have more green space amenities and less exposure to hazards. The separation of affluent and poor neighborhoods can be traced back to suburban sprawl, which increased air pollution, increased loss of green space to housing tracts and roads, and decreased walking and biking opportunities, resulting in health problems related to lack of exercise. Healthy and sustainable neighborhoods (chapter 25) is one way to achieve citywide sustainability. Green infrastructure is an important strategy for healthier neighborhoods, and cities are beginning to recognize environmental impacts in poorer neighborhoods by initiating more programs for recreation-based parks and natural green spaces.

The final chapters collectively discuss the concept of designing nature to suit human needs. Designing nature is based on the notion that landscape aesthetics—the visual quality of green spaces—can play an important role in sustainable cities by influencing stewardship practices for green spaces in urban areas. Chapter 26 describes how landscape professionals design nature and create urban landscapes, including the role of preferences for and perceptions of landscape aesthetics that impact ecological health and cultural experiences. Residential landscapes (chapter 27) are presented to discuss the notion of private property rights and the role of ordinances and laws for environmental protection and enhancement. Residential landscapes are significant green spaces in urban areas that present unique challenges as privately owned property. Stewardship depends on the homeowner's environmental ethic, horticultural knowledge, and desire for a pleasing landscape, which is typically influenced by neighborhood norms and professional designers. The role of landscape and planning professions as facilitators and decision makers is important for management of urban green spaces. More recently, experts in urban ecology and natural sciences have created social marketing programs and other incentives to encourage sound environmental practices and behaviors (chapter 28). Landscape certification programs are one incentive used by designers and policy makers to encourage science-based stewardship practices. Certification programs (chapter 29) are also often used in strategies, policies, and laws, and in social marketing campaigns to promote social awareness of the importance of protecting the environment. Finally, chapter 30 discusses the future of urban ecology, including the need to rethink many of our political, social, cultural, and economic dynamics in urban areas, especially in the context of environmental risks to people and nature with climate change impacts, and the promising ideas and programs for creating healthier sustainable cities.

Current examples of investigations and research into the future of urban ecology can be seen in two major metropolitan areas in the United States. Baltimore and Phoenix have been the sites of long-term urban ecology studies to investigate the ecological, cultural, and economic forces in cities, including many of the issues discussed in the chapters of Part III. Funded by the National Science Foundation, the Baltimore and Central Arizona-Phoenix Long-term Ecological Research projects, commonly called the Baltimore Ecosystem Study (BES) and the Central Arizona–Phoenix Ecosystem Study (CAP LTER), began in 1997; BES completed data collection in 2018 and CAP LTER will continue until 2021. The BES studies have focused on watershed biogeochemistry, ecological communities, and human environmental perceptions and behaviors. The goal of BES studies was to advance the understanding of urban areas as an important and novel ecosystem type, by considering three things: urban structure, function, and people. Objectives included determining the spatial and temporal structures of ecological, physical, and socioeconomic factors and the movement of energy, materials, and populations in the urban ecosystem, as well as investigating the choices made by people and their organizations that affect the urban ecosystem. Some of the key findings and products include the development of new theories and methods for describing urban ecosystems, resulting in a new urban-systems science which is now being used around the globe as a component of sustainability science for urban areas. The study has also transformed the long-held assumption of low urban biodiversity, discovering that biological communities in cities are diverse and have significant effects on water, energy, carbon, and nutrient fluxes. The research projects have also provided new information on the influence of environmental knowledge, values, behaviors, and perceptions of the citizens on ecosystem function and how ecosystem changes affect social and cultural concerns such as physical activity, perception of neighborhood desirability, and social cohesion.

The BES and CAP LTER projects have collaborated on data collection and sharing with similar survey platforms to produce direct comparisons between cities and comparisons with four other cities: Boston, Miami, Los Angeles, and Minneapolis-St. Paul. The projects have pioneered new theories and methods for describing urban ecosystems, established long-term urban data sets for watershed hydrology and biogeochemistry, and described long-term changes to urban biodiversity including with plants, birds, and soil fauna. They have also developed and applied long-term urban social survey instruments and worked directly with educators, schools, government agencies, communities, neighborhoods, and nongovernmental organizations to bring science into the community with strategies to improve environmental quality and human health and well-being. Both projects are currently working on education and outreach for broader impacts, including a Pathways to Environmental Science Literacy project in schools, data sharing with municipalities working on water quality regulations, and further partnerships and shared funding with universities and other government organizations (Central Arizona–Phoenix LTER 2019; Baltimore Ecosystem Study n.d.).

15

History of Nature in Cities

Nature has influenced the form and function of cities from the time of the first human settlements, where geology, hydrology, and vegetation helped create the infrastructure needed for habitation, commerce, and trade. The significance of urban ecological history to modern urban ecology lies in the generations of historical, cultural, and environmental contexts of the city that create a new paradigm for urban nature as sustainable landscapes for the future. In the age of the Anthropocene we have acknowledged that humans shape nature, but our heightened concern about the loss of urban nature gives us a reason to study how cities historically adapted to environmental changes (Girot 2016, 6–7). In urban planning the term "landscape" generally refers to the entire built environment, including natural elements and the tangible and intangible social and cultural qualities. Historic urban landscapes are treated one of two ways in the city planning process: (1) as a conservation project if a landscape is of significant value to the identity and character of a city; or (2) as a green space that provides historic context that may be conserved, adapted, or referenced as cities evolve. For planners the challenge for treatment decisions has been lack of an official definition and a description of the characteristics of historic urban landscapes. In 2011 the United Nations Educational, Scientific, and Cultural Organization (UNESCO) finalized a definition in its *Recommendation on the Historic Urban Landscape* as "the urban area understood as the result of a historic layering of cultural and natural values and attributes . . . to include the broader urban context and geographical setting." The geographical setting and broader context include topography, geomorphology, hydrology, forests, natural features, and open spaces and gardens. The recommendations include the need to better integrate urban heritage conservation strategies within the larger goals of sustainable development (UNESCO 2011, 3).

Traditionally anthropologists and archaeologists have engaged in the study of historic cultures and societies, but more recently environmental

historians (how humans interact with the natural world over time), eth-
nobotanists (how humans use and perceive plants), ethnoecologists (how
people perceive and manipulate their environment), and urban ethnohy-
drologists (how people perceive urban water quality and water management) have begun to study the historic implications in their fields. Other
traditional environmental scientists such as landscape and plant ecologists
and climate ecologists are also involved in the study of historic places.

Early historical records of nature and green spaces in cities almost exclusively feature gardens or parks because they were (and still are) the
defining green space in densely built cities. Our understanding of the cultural perceptions of nature in historic cities comes from drawings and murals depicting stylized layouts of private gardens and landscape paintings
featuring idealized picturesque city parks, which may not be an entirely
accurate historic record, but nonetheless illustrates the aesthetic ideas
that gave form to urban parks and nature in cities. Although many green
spaces exist in cities, including wetlands, floodplains, road rights-of-way,
and remnant forest patches, this chapter focuses on human-made gardens
and parks as the primary historic natural landscape that is significant for
sustainable development in cities today. Three historical phenomena that
were most influential for green spaces and ecology in modern cities include the concept that humans created a different version of nature by collecting and arranging plants in gardens and parks, the invention and use of
irrigation, and changing perceptions about nature that shaped landscape
styles and recognized the need for nature in cities. Although few citizen
science projects are related to history and urban areas, one project is em-
ploying volunteer papyrologists, who study literature preserved on paper
made from the papyrus plant, to help translate text on ancient papyrus
sheets.

What: Evolution of Gardens, Irrigation, and Nature Attitudes in Urban Settlements

The earliest gardens had simple wood fence structures to protect domesti-
cated plants and animals. These walled gardens were the first intentionally
designed landscapes in the "urban" setting of small villages of the Neolithic
period (4000–3000 BCE). The growth of cities and invention of irrigation
changed agricultural practices, creating new wealth and status levels for
some that led to a new concept of private walled gardens for leisure. The
murals and paintings from ancient Egyptian, Babylonian, Assyrian, and

FIGURE 15.1. Marmaris, Turkey: walled garden of the Marmaris Castle, first constructed in 3000 BC and rebuilt in 1522 by Ottoman Sultan Süleyman.

Persian gardens (4000–500 BC), and later from the Roman Empire (510 BC–AD 476), depict elaborate walled gardens with fruit trees, flowering plants, wildlife, and fish in ponds arranged in a stylized version of nature established by the linear shape of the irrigation runnels. Over time, walled gardens grew to include collections of plants from exotic countries to display the wealth and power of the owners. As plant explorers traveled the trade routes to the Far East and Europe, the concept of the walled garden spread from the Middle East to China, India, and Western Europe (Girot 2016, 42). In the Middle Ages (AD 476–1350), walled gardens were used in monastery cloisters within castle walls. The wealthy enjoyed walled gardens in Islamic and Moorish settlements (AD 700–1400), and defense structures of the same period, including fortresses and castles, often had planted courtyards within their walls (fig. 15.1). The walled concept from the Roman peristyle gardens and Christian cloisters spread into North Africa and Spain. By the late twelfth century, walled gardens eventually made their way, via Spanish explorers, through South America to the Christian missions in the American West.

Gardens and parks in the United States were influenced primarily by the European garden styles brought by settlers and explorers from

England and Spain. Similar to the original walled gardens, American colonists (1700s) had small fenced gardens in front yards for growing food and
keeping animals. At the same time, the Age of Enlightenment in England
(1700–1800) gave rise to science and changed the focus of private gardens
from elaborate designs (fig. 15.2) to collections of plant materials. With the
introduction of new plants and a new genre of landscape paintings that
emphasized an idealized nature, there was a shift in aesthetic preferences
and attitudes about nature among the upper class, who built estate gardens to reflect this new nature. Eventually the industrial revolution in U.S.
and European cities created intolerable urban conditions, and the social
reform movements began to advocate for public green spaces, establishing the concept of a public park for everyone. At the beginning of the
nineteenth century a shift occurred from aesthetics to the functional use
of parks for recreation for working class people and playgrounds for children. In the United States, legislation for zoning that separated work and
living spaces included calls for more park spaces as the means to an ideal
urban life (Sadeghian and Vardanyan 2015, 122). Central Park in New York
City was one of the first of many parks designed in cities in the United
States, adding a new landscape concept to the design vocabulary for cities:
the urban park.

FIGURE 15.2. London, England: an elaborate design in the historic sunken garden at
Hampton Court Palace.

FIGURE 15.3.
Central Italy:
large historic
irrigation run-
nels used for
irrigation of an
ancient grove of
olive trees.

The invention of irrigation made walled gardens, yards, and parks possible. The concept of transporting water to grow plants began as a simple
network of runnels (fig. 15.3) that relied on gravity to move water and
evolved into the sophisticated systems that keep our cities green today.
Irrigation made possible the first settlements and early city-states, today
known as the "hydraulic civilizations" that first appeared in the Euphrates
delta at the tip of the Fertile Crescent. Irrigation canals in grid patterns
gave form to the square geometry of walled gardens where they were first
used. Gardens were built where the natural slope of the land could be ter-
raced into quadrants so water could flow from one quadrant to the next
(Girot 2016, 45, 49). Around 540 BC the city of Ur had an elaborate network of canals and levees running perpendicular to the Euphrates River
for a constant source of water that ensured successful harvests. Those with

wealth, the merchants and owners of the best plots, built lavish walled gardens, which began the creation of a society differentiated by wealth. With the promise of a constant food source, the city flourished and the need to manage irrigation for crops and gardens gave birth to the notational systems that helped develop time, geometry, mathematics, and writing. Surveying and crop records on clay cuneiform tablets were some of the earliest forms of writing. The Sumerians also conceived the concept of time as a division by 60 (the sexagesimal system) as a means to monitor water distribution, and they developed the mathematical measurement of surface geometry by triangulation, developing a land surveying system we still use today (Girot 2016, 53–54). Other irrigation systems included qanats (vertical and horizontal well systems located on the sides of hills) from Persia, the Roman aqueducts, Muslim waterwheels, and windmill pumps from the Muslims and Dutch. By 1800, irrigated acreage worldwide had reached about 20 million acres; today it is an estimated 700 million acres (New World Encyclopedia 2018).

Today city parks and irrigation are commonplace in urban areas, but designers are beginning to rethink irrigation practices to conserve water and reimagine city parks as valuable multifunctional green spaces to help mitigate climate change stressors. Urban parks are used for flood control, recreation, relaxation, wildlife habitat, plant conservation, and to moderate the heat-island effect. They also provide cities with unique identities with new ecology-based aesthetics and a new approach where historic context (fig. 15.4), development, and appreciation for nature can interact to increase sustainability in urban areas (Gobster 2007, 97).

So What: Documenting Urban Areas' Natural History

Landscape architect Ian McHarg, author of the pioneering urban design book *Design with Nature* (1969), believed we should visualize the urban landscape as interconnected layers of history to understand the social-environmental relationships of people and nature. Using his multidisciplinary approach with layers of inventory information from above, below, and within the urban ground, he incorporated science with nonecological issues such as history and policy, showing the causal relationships of conditions in the built environment (Orff 2016, 10). When urban planners and designers began to use this technique, a new appreciation for the historical context of the city emerged, and recognizing history in urban development

FIGURE 15.4. Paris, France: a modern-day stormwater irrigation runnel in a park that replicates the historic runnels for watering trees.

began to take three different approaches. One is the conservation and restoration of human-made historic green spaces, such as parks, cemeteries, and streetscapes. The second approach is to use the historic ecology of the area, such as geologic formations, original hydrology patterns, and native plant communities to create, or re-create, urban green spaces; and the third approach is to layer historic ecology, the present-day urban ecology, historic and modern-day cultural needs, and the urban structure to create spaces that strengthen the city identity and improve modern urban life. In some cases, taking the long view over short-term fixes, such as choosing not to build somewhere, or choosing to build differently, is seen as the key to a city's survival. Global and national programs are available to cities to document and characterize historic landscapes in reports intended to guide city planners and designers.

The National Park Service Heritage Documentation Program launched the Historic American Landscapes Survey (HALS) in 2000 to record historic landscapes in the United States and its territories. The process uses measured drawings, interpretive drawings, written histories, and black-and-white and color photographs to document historic landscapes, including urban and rural, agricultural and industrial, and designed and vernacular landscapes. Landscape types include small gardens, estate gardens, cemeteries, suburbs, abandoned settlements, quarries, farms, and nuclear test sites. The Park Service determined that historic landscapes are important to national, regional, and local identity and are too often affected by the forces of nature, development, and neglect, and for the benefit of future generations the sites must be documented. Guidelines for historical reports include historical information and physical information, including landscape character and use, overall description and condition, and landscape features and conditions. The documentation is intended to convey the significance of the site and the context in which it was created and evolved (Robinson, Vernon, and Lavoie 2005, 5).

Several countries in Europe, including England, Scotland, and Ireland, have official programs using the concept of historic landscape characterization to increase their understanding of designed landscapes by looking at the essential qualities and character of a place. Besides ornamental landscapes and archaeological features, the programs look at all types of human-made natural features such as managed woodlands. The goal is to manage and protect the landscape within the planning and design of current projects. Objectives include documenting the existing situation, identifying gaps needing more investigating, enabling participation by local residents, and providing reports to planning and land use authorities. Landscapes are characterized by the way they were formed, earlier and current land uses, and their physical appearance using historic and current maps and aerial photographs. When reviewing proposals for current projects, planning authorities are required to take into account the historic dimensions of the landscape (Historic England 2020).

Now What: Historic Landscapes in Today's Cities

Several different strategies are being used to recognize the historical context of a city; one approach is to consider the large-scale historic context of the entire city and surrounding area, another approach is to concentrate on the most historically significant small-scale spaces with conservation and restoration projects. UNESCO believes that historic context is important for new development in ever-evolving cities, recommending the layering of cultural and natural values over time, including geomorphology, topography, open spaces, hydrology, and infrastructure to sustain and enhance the identity of cities. UNESCO's historic urban landscape approach includes:

- a full assessment of the city's natural, cultural, and human resources;
- using participatory planning to decide on conservation actions;
- assessing vulnerability to climate change and socioeconomic pressures;
- integrating urban heritage values in city development;
- prioritizing policies for conservation and development;
- establishing public-private partnerships; and
- developing mechanisms for coordination.

The approach advocates that historic context and new development can interact to support each other and integrate conservation and social and economic development. The goal is to increase sustainability by integrating environmental, socio-cultural, and economic concerns in the planning and design of urban development, recognizing that a city is a continuum in time and space. Examples of projects recognized by UNESCO include a 10 hectare (25 acre) urban recreation park that includes 5 kilometers (3 miles) of vehicle-free space along the banks of the Rhone River in Lyon, France. The park is designed for daily use and to support special historic events. Another project focuses on protecting the canals of Amsterdam for visitors and residents alike. The unique approach to the protection program is the recognition that the strength of the canal district is the residents of the district, and the importance of finding balance between residents and visitors (UNESCO 2013).

The most common urban green spaces targeted for conservation are usually natural areas, such as stream corridors, wooded areas, beaches, and unusual geological formations. Human-made historic landscapes such as parks, cemeteries, famous streets and promenades, and water features such as artificial lakes are usually targeted for restoration. Designed landscapes, such as gardens, present challenges because of the changes in plant material over time. Restoration means restoring a garden to its character at a previous point in history and considering the current and future use of the space. Restoration requires research and documentation of the garden by practitioners with knowledge of landscape archaeology, history, landscape design, and horticulture. Many nations have statutory protections for registered parks and gardens and organizations such as historic garden societies, garden conservancies, and national trusts for historic preservation.

Citizen Science Program: Uncovering "Ancient Lives" with Citizen Science

SciStarter has launched the Ancient Lives Project, inviting citizen scientists to transcribe ancient papyrus texts from Greco-Roman Egypt. In 1897 the excavation of the ancient Egyptian city of Oxyrhynchus, the "City of the Sharp-Nosed Fish," uncovered hundreds of thousands of papyrus fragments that had been preserved in the arid climate. The texts, which date from third century BC to seventh century AD, include works of Plato and St. Matthew's Gospel along with more mundane texts such as marriage certificates, personal letters, land leases, and trade transactions that have given scholars the opportunity to study the working lives of people who lived centuries ago. The papyrus collection is currently at Oxford University, where only a small percentage have been transcribed from their original ancient languages. For the Ancient Lives Project, papyrologists have collaborated with computer scientists to expedite the process of reconstructing the documents and transcribing them. Volunteer papyrologists (people with no expertise in transcribing papyrus documents) are helping to transcribe digitized texts that have been uploaded in a database. The project website includes tools to help with the transcription, and since the summer of 2011 the project has logged more than 1.5 million transcriptions and they are still working on millions more (Modi 2014).

Project: Town Branch Commons

SCAPE is a landscape architecture firm that engages in projects with communities to bring a holistic, activist approach to working with urban nature. A recent proposed project with the city of Lexington, Kentucky is an example of a design strategy to reconnect with the historic urban nature. A buried stream channel that historically served as a canal for waste, sewer, and water became the catalyst for a new water-based public realm using the region's karst geology as inspiration. The highly porous karst geology creates unexpected water flow patterns, a characteristic that will be used to create a karst identity with a series of pools, fountains, bioswales, and filter gardens to expose the underground stream. The design team used a series of section drawings to study the natural karst hydrology and another series of section drawings to illustrate how the urban interpretation of the karst hydrology would work at Town Branch Commons. The design centers around four actions: (1) reveal, by daylighting the stream and creating a series of flood-adapted recreational and ecological rooms; (2) clean, using water filtration gardens in a pedestrian streetscape to clean waters before they enter the culvert below; (3) carve, building a public event plaza with constructed karst windows filled with water; and (4) connect, using a small headwater stream to create a "blue-street" link between two historically divided neighborhoods and create pedestrian and bicycle links with a 20 mile regional trail. The proposal was a winning entry in the Town Branch Commons competition and has evolved into a strategic planning effort and hydrologic feasibility study with phased construction funded by city, federal, and private dollars (Orff 2016, 27–44).

16

Mitigating the Impact
of Human Activities

Many global, national, and community programs and initiatives are available to help cities become more sustainable, reduce human impacts, and protect the urban environment. Strategies include economic, political, ecological, and cultural programs for sustainability. Most of these programs rely on policy and regulation, design strategies, planning processes, and implementation strategies to achieve improvement goals. Examples include resilient city planning practices to protect against rising oceans and climate change and investing in infrastructure for water recycling and desalination. Economic strategies include adopting renewable energy technology and creating pedestrian-oriented streets to reduce dependence on automobiles. Policy includes zoning practices to reduce urban growth and sprawl and adopting new urban forms, such as edge cities, to increase diversity and density. Ecological approaches include improving urban agriculture opportunities and using native ecologies to support urban design principles. Development processes include designing architecture and buildings to be self-sustaining and biophilic, meaning they provide a healthy environment for people (Erikson 2016). These large-scale goals and strategies can be accomplished in several ways, but they all rely on citizens embracing change and accepting some responsibility by exhibiting pro-environmental behaviors to mitigate the environmental impacts. Urban development is clearly destructive to the natural environment; loss or damage to vegetation (fig. 16.1) and disturbance of the natural soil profile during site clearing and construction is only the beginning of the environmental degradation in cities. The ongoing use of harmful maintenance practices is a commonly cited human activity that contributes to environmental problems, including chemical pollution from fertilizer and pest

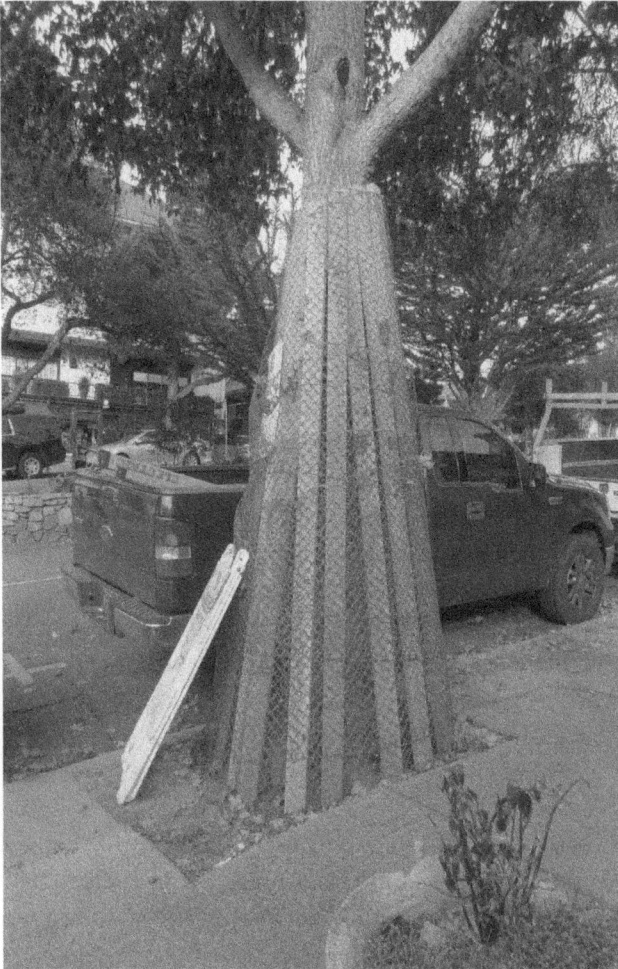

FIGURE 16.1. Carmel, California: protecting street trees during construction with barricades.

control applications, spread of plant diseases from poor equipment maintenance, and spread of invasives through poor management activities.

Efforts are being made to mitigate management and maintenance impacts at several levels. Scientists at many university extension programs have developed management practices for a variety of landscapes. The most common research areas for management activities include developing new technologies and improved plant cultivars and developing multilayered maintenance practices. For example, irrigation experts are testing new technology such as soil moisture sensors and smart controllers to reduce water use and automate irrigation schedules. Turf scientists are

developing new sod varieties that are drought and pest tolerant with limited growth heights to reduce mowing. Seed biologists and plant breeders
are developing new cultivars that need fewer inputs and sterile cultivars to
reduce invasive plant spread. Different Integrated Pest Management (IPM)
strategies are being developed by weed scientists to target weeds without
harming other plants and reduce chemical inputs. Citizen science projects are also interested in maintenance activities. Dozens of water quality
testing projects in urban streams focus on pollutants from pesticides and
fertilizer; soil sampling projects are testing for chemical contamination,
and weed spotter projects are helping to map weed spread and abundance.

Citizens can also make a difference by focusing on pro-environmental
maintenance practices in residential yards and private urban green spaces.
Programs for site management based on best practices and behaviors include Integrated Pest Management (IPM), Integrated Vegetation Management (IVM), Best Management Practices for Preventing the Spread of Invasive Plants (BMPIP), and Green Industries Best Management Practices
(GIBMPs). The Audubon Cooperative Sanctuary Program (ACSP) for golf
courses and the California Invasive Plant Council BMPIPs are examples
of comprehensive maintenance and prevention programs that incorporate
IPM, IVM, and BMPIP practices.

Golf courses are common urban green spaces that have been highly
criticized for a variety of environmental problems attributed to maintenance practices, and as a result the golf industry has responded with programs that use alternative practices to make them more environmentally
friendly. Public transportation and utility corridors are also often cited
for poor maintenance activities and as conduits for movement of invasive
plants to and from the city.

What: Analyzing Impacts and Mitigation Opportunities

The lessons learned from the efforts of the golf industry to reduce environmental impact, and public land managers to control invasive species,
can be applied to many urban green spaces, including a typical residential
neighborhood lot. ACSP program guidelines are presented in this chapter
to demonstrate an approach that can be adapted for any private or public
urban green space at a range of scales. The average golf course is about
150 acres, the equivalent space of 600 homes in a development of standard quarter-acre lots. Although golfing has lost some popularity among
millennials (the National Golf Foundation estimates about 205 courses

closed in 2017 due to oversupply and under demand), the game is not likely to disappear soon (Biron 2018). Approximately 14,800 golf courses exist in the United States and most of them are in the southern sunbelt area where water is scarce, which has generated criticism about the use of irrigation water to maintain vast amounts of turf. Negative environmental impacts of golf courses include loss of natural habitat and groundwater pollution caused by fertilizers and pesticides. However, a well-managed, well-planned golf course can provide soil protection, water filtering, bio-diversity conservation, and wildlife habitat (Audubon International ACSP 2019). In response to the criticism, the golf industry has implemented several strategies to reduce environmental impact and change the public's perception of golf courses, the most prominent strategy being the Audu-bon Cooperative Sanctuary Program (ACSP).

Golf courses, including private, public, and municipal courses, can join the ACSP and develop an environmental plan that fits their unique situa-tion. ACSP provides a Site Assessment and Environmental Planning form to guide the process and educational material on topics such as wildlife and habitat management, environmental planning, reducing chemical use, water conservation, water quality management, and outreach and edu-cation. When a course implements and documents the use of practices in these areas, it is eligible for designation as a Certified Audubon Co-operative Sanctuary. The goal is to improve environmental function and community relations, save money on maintenance practices, and conserve environmental resources. Some of the long-term results that are generally found in the program include improved quality of the natural resources on the course, such as better water quality in ponds, healthier plants and turf, and more species of wildlife living in and around the course. Courses also save money with reduced costs for energy, water, fertilizer, pesticides, and labor. Using best practices has also proven to reduce workers' expo-sure and risks associated with standard chemical treatments. To prove that implemented practices are being effective, participating courses must pro-vide photographs with a labeled map of the course and their IPM records. They must also provide the results of required water quality tests, a before and after wildlife inventory, and samples of the educational materials they have developed. To ensure the practices are science-based and effective, program members must also create an advisory group of staff members, golfers, and local experts. Experts can be from university extension pro-grams, water management districts, local environmental organizations, or government environmental agency representatives. Recertification is required every three years to ensure members are following the required standards (Audubon International ACSP 2019).

Public transportation and utility corridors are another example of urban green spaces that have been blamed for detrimental environmental impacts due to poor management practices. The primary issue is the relatively easy movement of plants into and out of the city by human activity. Plant choices and management practices are a factor in the spread of non-native plants from urban areas to surrounding natural areas where they can become invasive and disrupt native plant communities in natural areas. While non-native plants are easily moved from urban gardens to wild areas, it does not necessarily mean they will become invasive, but the high number of ornamental plants purchased and installed in urban green spaces increases the probability that some will adapt to wild habitats and spread, disrupting native biodiversity. Non-native plants, also called exotic or alien plants, are those introduced from a different part of the country or world, either accidentally or on purpose. Many are used in the ornamental landscape industry and do not cause environmental problems; for example, Florida Exotic Pest Plant Council (FLEPPC) reports that of the nearly 1,400 exotic species introduced into Florida, 11% have become established outside of cultivation and are considered invasive (UF/IFAS CAIP 2018). Although this is a small number, many are problematic because of their ability to rapidly take over large areas of land or water, displacing both non-invasive, non-native and native plant species. Invasive species are described as non-native species whose introduction does, or is likely to cause, economic or environmental harm to humans, animals, or plants.

Invasive plants tend to be habitat generalists with high adaptability to new habitats that reproduce easily and outcompete other plants, and they typically are not affected by native pests and diseases. They generally appear first in urban areas as seeds or weeds that come from imported nursery plants and soils. Travelers to other countries can unknowingly, or purposely, carry plants and seeds for aquarium or house plants, and sometimes shipping containers carry seed hitchhikers. Invasive plants are also spread naturally by climatic events, and by migratory birds and fauna (EDD Maps 2019). Experts are investigating the characteristics of plants and mechanisms by which non-native plants become invasive, including propagule pressure, plant traits, and habitat conditions. The challenge is to predict which non-native plants have the potential to become invasive. Citizens can help prevent the spread of invasives by purchasing plants at nurseries that don't sell restricted or known invasive species and by considering use of native plant species (UF/IFAS CAIP 2018). It is important to note that native plants are sometimes erroneously labeled invasive if they grow aggressively. Native plants can become aggressive growers under certain conditions, especially in urban green spaces that provide unique habitats that favor their spread. When native plants are able to take advantage of disturbed soil and outcompete other plants, they are usually referred to as opportunistic native plants rather than invasive (Ecological Landscape Alliance 2019).

So What: Reducing Environmental Impacts of Maintenance Practices

IPM and IVM practices can be adapted to any urban green space to reduce the impacts of standard chemical-based pest and weed management. The IPM Institute of North America defines IPM as a "science-based decision-making process that identifies and reduces risks from pests and pest management related strategies. The goal is to provide an effective, comprehensive, low-risk approach to protect resources and people from pests" (IPM 2019). A nature-focused definition for IPM that applies to green spaces describes IPM as an ecosystem-based strategy that focuses on long-term prevention of pests and their damage using biological control, habitat manipulation, modifications of cultural practices, and resistant varieties. Integrated Vegetation Management (IVM) practices parallel IPM strategies using biological, cultural, mechanical, and chemical control to manage unwanted vegetation, including invasives and weeds (United States Botanic Garden 2014, 25). IPM and IVM pests are described as (1) plants (weeds or invasives); (2) vertebrates, such as birds or rodents; (3) invertebrates, including insects and snails; and (4) nematodes and pathogens, such as bacteria or viruses (UC IPM 2019). The main concept of IPM and IVM is using strategies that target the pest, create unfavorable conditions for it, and limit impact on other organisms and the environment. Applications of chemical pesticides are the last resort. Key practices for ecosystem IPM and IVM include inspection and monitoring to accurately diagnose pest problems and developing the best combined management approach by using several methods. Biological controls use natural enemies such as predators, parasites, pathogens, and competitors to control pests. Cultural controls reduce pest establishment by limiting reproduction, dispersal, and survival. Mechanical and physical controls kill pests directly or make the environment unsuitable; examples include traps for animals, mulches for weeds, and screens for insects. Chemical control uses selective pesticides such as bait stations or spot-spraying of weeds (UC IPM 2019). Plant-friendly strategies also include removing pests by hand, using water showers or baths, and using organic oils. EPA "reduced risk" pesticides include biopesticides, which are naturally occurring substances or microorganisms produced by plants that contain genetic material introduced to control pests. Most new pesticides meet the new criteria (since 2003) and reduce contamination of groundwater and surface water (IPM 2019).

IVM strategies differ slightly from IPM. While they both address undesirable species, IVM also includes planning and promoting the use of desired plant communities (fig. 16.2) that are stable and resist invasion by undesirable (usually invasive) plants. Recommendations for control of undesirable plants include using herbicides approved by the EPA in a focused, selective manner, particularly for invasives that are difficult to manage mechanically. Biological controls are also recommended to reduce plant populations to manageable levels, and cultural controls are used to create conditions to introduce desirable species. Manual or mechanical controls such as mowing, hand-pulling and prescribed burns are also methods that can be combined with herbicides or biological controls (United States Botanic Garden 2014, 25–26). Prevention, including eradicating invasives using early detection and rapid response (EDRR) on small populations in corridors before they reach wildlands, is a key aspect of invasive plant management. The California Invasive Plant Council has developed a set of BMPs for wildlands and transportation and utility corridors that can be adapted to other urban green spaces. General BMPs for corridors include evaluating risks and scouting for invasive plants before beginning maintenance activities and planning travel routes to avoid infested areas (Cal-IPC 2012a, 1–2, 4, 7–8). Planning BMPs include (1) adopting an official policy to prevent invasive introduction and spread; (2) integrating invasive plant BMPs into design and construction planning documents; (3) coordinating prevention efforts with adjacent property owners, and (4) developing a monitoring plan. Site activity BMPs include recommendations for project materials, such as using weed-free materials like clean mulch, using management methods that favor desired plants, and when landscaping, using local plant materials to revegetate as soon as possible before invasives can establish. Activities also include inspecting and cleaning tools, equipment, vehicles, and clothing before entering and leaving the worksite. Designated cleaning and waste disposal areas are also used to contain invasive materials. Prevention methods include minimizing soil

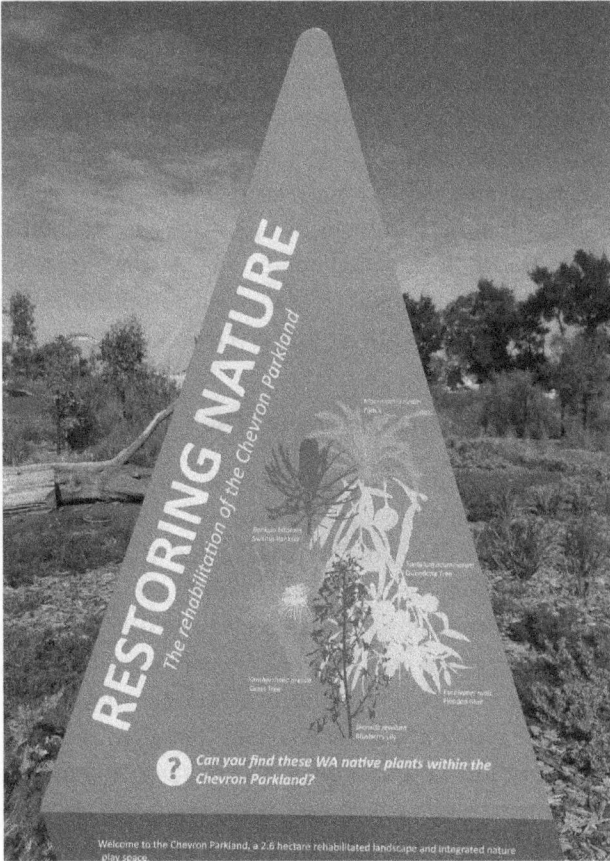

FIGURE 16.2. Perth, Australia: restoration of Chevron Parkland with native plants to promote the use of desired plant communities.

disturbance and erosion to discourage invasive establishment and retaining existing desirable vegetation (Cal-IPC 2012b, 7–8).

Now What: The Future of IPM and New Science on Invasives

Understanding why some plants become invasive and others do not is a nascent area of plant science. New research from the University of Vermont provides some insight into predicting the likelihood of a plant becoming invasive. Plant height may be one of the biological traits that give some non-native plants an advantage over other native plants in certain plant communities. By comparing traits of native and non-native plants in a range of plant communities, differences were found in the non-native plants that did not spread aggressively and those that did spread rapidly and became invasive. The non-invasive plants shared similar traits with the native plant community, while the traits of invasive species were similar enough to allow them to live in the same habitat but different enough to allow them to thrive. For example, some invasive plants were taller than the native species, which could give them better access to light to outcompete native plants. The findings support a novel theory of invasion called the edge-of-trait-space model that suggests that non-native plants can coexist in a native community but can invade if they have slightly different adaptations to the environmental conditions. A single, easy to measure trait, such as height, could predict if plants will become invasive, and eradication efforts could focus on those plants (Divisek et al. 2018).

Citizen Science: Environmental Archaeology at SERC

Citizen scientists are helping to uncover the traces of the ways humans used and abused the land surrounding an old house and farm at the Sellman Plantation, on the western shores of Chesapeake Bay. The house was part of a thriving plantation from the 1700s called Woodlawn and is now part of the Smithsonian Environmental Research Center (SERC). Teams of volunteers are overturning soil to look for evidence of how the land was used and managed. For more than 250 years the owners cleared land, tilled the soil, planted non-native plants, and dumped trash. They also used poor farming practices that contributed to erosion and unnatural land forms. Participants are examining soil on the site to address three major research concerns: erosion and sedimentation, species diversity, and coal usage. The goal is to learn about the relationship between the land and the people to study the ways that humans interacted with the environment in the past and how we could address similar problems today (Minogue 2012).

Case Study: ACSP Golf Course Case Studies

Case studies describing some of the strategies used by ACSP participants include a course in Harrison, Tennessee, that installed 45 bird nesting houses and a large planted area with 218 plants native to Tennessee. The course also set aside 40 acres as natural area to minimize maintenance and switched to a different species of turfgrass that reduced chemical use and lowered the budget from US$39,000 annually to US$8,000. The course received a state-sponsored award for their stewardship and commitment to sustainability. A course in Las Vegas, Nevada that was under increased public scrutiny regarding water use removed a 7,500 square foot area of sheep's fescue, a high water-use grass, and replanted with drought-tolerant native plants. Three hundred plants were installed, representing 18 species of desert plants. The desert garden has resulted in a savings of approximately half a million gallons of water a year and requires only 10% of the human-hours needed to maintain the fescue grass (Audubon International ACSP 2019).

17

The Legal Landscape

There are two sets of legal parameters that determine outcomes in our landscapes: environmental legislation and land use legislation. Another dimension that dictates basic precepts is related to jurisdictions. Constitutional and federal laws provide the basis for uniform treatment of issues at the national level. State legislation enables local governments to devise their own plans, regulations, and municipal codes, which usually contain the specificity necessary for proper implementation at the appropriate jurisdiction level. Environmental legislation is more commonly determined at the federal level because the environment does not recognize political boundaries, natural features usually transcend state lines, and environmental issues may affect entire cohorts of the population. Land use legislation is usually determined at the local level (municipality or county) because it establishes parameters that affect local territories, regulates issues of local concern, and may impact citizens at a neighborhood or community scale.

Environmental legislation, as with all laws, varies with the legal framework of a given country. In the United States, where Common Law is the legal system, legislation is enacted by Congress, and the Supreme Court establishes precedents when any aspect of the law is challenged. In countries where Common Law is the system, cases decided by judges can establish precedents that then become law and establish the parameters against which future cases will be judged and ruled. Canada and Australia are also Common Law countries, following the tradition of all British colonies; however, differently from the United States, both are still part of the Commonwealth, and thus their legal framework is not the same as that of the United States. In addition, in Canada, the Province of Quebec follows Civil Law, after the French tradition and system. Other countries, particularly those colonized by the Spanish and the Portuguese, follow a Civil Law system, also known as Roman Law, which is yet another legal framework that dictates how environmental legislation is passed, followed, and enforced. In countries where Civil Law is the legal system, there are no legal precedents that can change a law or code in place; rather, there needs to be a legislative act to change the current codes. A Civil Code serves as the basis for all legal decisions in countries where the codification of private law is founded on it.

Because land use codes are for the most part specific to the local sphere, we will limit the discussion to American land use. From the time our forefathers enacted the first pieces of legislation, private property rights have been considered sacred. All the land use legislation in the United States recognizes that property owners have complete control over their land and only under special circumstances is that right meddled with. The most common legal mechanism applied in special circumstances is eminent domain. Takings law is an entire specialization, and explaining it would be beyond the scope of this book.

In global terms, multicountry agreements, such as the Kyoto Protocol and the Paris Accord to name but a couple of the most recent ones, are the vehicle that approximates legislation the closest, although international agreements do not have the same power and accountability of the legal framework of a single country.

What: The Legal Sphere

In the United States, federal laws are submitted as Bills to Congress and, once approved, they become Law. Laws enacted by Congress are referred to as Statutes. Federal agencies are responsible for creating the regulations so that laws can be enforced; this process is known as rulemaking (United States Government 2019). For example, the Environmental Protection Agency (EPA) is responsible for detailing and enforcing all environmental legislation approved by the House of Representatives and the Senate. All federal laws are underpinned by the U.S. Constitution. The most significant environmental legislation enacted by the federal government includes the Solid Waste Disposal Act of 1965, the Air Quality Act of 1967 (precursor to the Clean Air Act of 1970), the National Environmental Policy Act of 1969, the Clean Air Act of 1970 (amended in 1990), and the Clean Water Act of 1972, which was based on the Federal Water Pollution Control Act of 1948 (EPA 2019a). The enactment of and amendments to these federal laws is a testimony to the environmental consciousness that became ubiquitous in the 1960s and 1970s, not only in the United States, but also in many countries in the world. In addition to environmental laws, the development of cities can be impacted by other federal laws that regulate the use of land. One of the federal laws that exemplifies that impact is the National Historic Preservation Act of 1966.

The 50 states that make up the United States, albeit part of a single union, are considered independent and thus have their own constitutions. The same legislative process that exists in the federal sphere is replicated at the state level. State legislatures approve laws that are subject to interpretation and state courts. The state in turn may use enabling legislation to delegate rulemaking power to local governments. The two most significant pieces of legislation proposed and developed by the Department of Commerce are the Standard State Zoning Enabling Act of 1922 and the Standard City Planning Enabling Act of 1928 (Bernard 1965). Enabling legislation is an important tool, especially for local communities, as it gives them the discretion necessary to enact zoning regulations, ordinances, and other rules that govern land use—such as subdivision controls and regulations—that impact every aspect of urban development. The four categories of enabling legislation considered "model laws" for cities are a planning enabling act, a subdivision enabling act, a mapped streets act, and a zoning enabling act (Bernard 1965, 634).

The first city in the United States to adopt zoning laws was New York in 1916 (Krueckeberg 1983), and cities in most states followed suit in the first half of the twentieth century. Urban renewal movements in the 1960s and environmental movements in the 1960s and 1970s gave rise to the first policies and subsequent land use and environmental legislation.

More recently, with the advent of private subdivisions, gated communities, and condominiums, homeowners associations (HOAs) are the organizations that establish the codes, covenants, by-laws, and regulations internal to developments and neighborhoods (fig. 17.1). HOAs are private governing bodies, usually constituted by developers and transferred to homeowners, responsible for maintaining and managing a residential development or neighborhood. In addition to complying with municipal or county laws as in any other development, residents of these private communities are required to follow rules that, for example, specify the use of common areas, such as open space and parking lots; architectural and maintenance standards, which impact the aesthetics of individual units and of the entire development; and occupancy limits, including the number of residents and pets in each unit (Grassmick 2002; Weinstein 2005).

FIGURE 17.1. Curitiba, Brazil: gated communities are physically and administratively segregated from the rest of the city.

HOA rulebooks are also known as HOA Covenants, Conditions, and Restrictions (CC&Rs).

All legal spheres are interconnected. If federal laws do not offer adequate protection against disputes between states, the courts decide. A cross-boundary issue bound to become a classic in law literature is the dispute between Georgia and Florida over the waters from rivers in the Apalachicola-Chattahoochee-Flint watershed. Florida has been trying for decades to limit Georgia's usage of water from this river basin, because the reduced flows caused the Apalachicola oyster fishery to collapse. Nonetheless, the circuit court judge recommended that the Supreme Court uphold the right of the state of Georgia to continue drawing water from the river (Lam 2019). In addition to state enabling acts that give rulemaking power to municipalities, some states have laws that protect the rights of homeowners within municipalities so that the HOAs do not infringe their private rights. For example, 19 states protect the right of homeowners to solar-dry their clothes, that is, to use clotheslines (Weinstein 2005; Howland 2012). For the most part, cross-jurisdictional disagreements have to be decided in court.

So What: Determining the Impact of Legal Frameworks

The legislative aspect of environmental protection and other policies that impact nature in cities is, in a way, the least difficult dimension of the issue. Enforcement of laws, rules, and regulations is what validates the legal frameworks and what essentially protects nature and mitigates the impact of cities. In the last ten years or so, there has been increasing concern that both federal and state laws intended to protect the environment in the United States have not been properly enforced (NACWA 2009). The media has reported on lack of enforcement; however, the current political climate makes good stewardship very difficult. In addition, the agencies responsible for enforcing legislation and policies do not seem inclined to do so in a holistic manner, addressing outcomes rather than actions. For example, in 2009 the EPA created a Clean Water Enforcement Action Plan that, despite broad consultation, was narrow in scope and did not recognize the real issues affecting water pollution at the watershed level (fig.17.2).

At the local level, legal frameworks are impacted by more subjective criteria. There are about 250,000 HOAs in the United States, and an estimated 80–90 million people live in communities regulated by one of them

FIGURE 17.2. Newark, New Jersey: water pollution and eutrophication caused by runoff, particularly along roadways.

(Grassmick 2002; Weinstein 2005). The growth of this type of development accelerated in the 2000s, with four out of every five housing starts being part of a homeowners association (Weinstein 2005). HOA regulations are a relic from the Napoleonic Code and, even though the United States operates under Common Law, which lacks a clear statutory framework for HOAs, it is still possible to have this type of covenant (Grassmick 2002). HOA rules vary from place to place; however, for the most part, they are more restrictive than municipal codes, and membership is not voluntary. The restrictions imposed on owners living in these communities are supposed to enhance quality of life; however, some have attempted to be exclusionary and ended up in court, with decisions varying considerably from state to state (Grassmick 2002). An interesting fact is that, with the advent of private subdivisions, gated communities, and condominiums, residents have grown accustomed to accepting that these codes and rules curtail their freedom in ways that would be objectionable in open communities. There seems to be more acceptance of private regulation of aesthetics and behavior than would be possible through public regulation.

Now What: Legal Models for Sustainable Cities

Very few people today would dispute the notion that the health and survival of our planet is a global issue. And with two-thirds of the global population living in cities, urban solutions are global solutions. Although we will have to continue relying on nations to enact policies and implement legislation to support making our cities sustainable, international organizations can influence individual countries and, to a certain extent, mandate compliance with guidelines that protect the entire planet. International treaties and multilateral agreements are but one way to influence individual countries and promulgate directives that can be adopted. Although international agreements, including treaties, declarations, charters, and letters are a good foundation for global sustainable development, their effect can be felt in cities only if institutional frameworks are established at the national and local levels.

The way national and local leaders and policy makers think about development has gradually changed with continued pressure from international organizations. Most countries today, especially those in the developing world, count on the financial and expert assistance of international aid and development agencies that impose holistic practices. These agencies have adapted their approach to development based on the experiences of the past 50 years, and today they include provisions in their contracts that range from welfare community programs to environmental requisites. The United Nations has spearheaded most, if not all, multicountry initiatives to address the current global environmental crisis. For example, the Intergovernmental Panel on Climate Change (IPCC) is a UN body created in 1988 and dedicated to addressing climate change issues from a scientific perspective, analyzing impacts and providing mitigation and adaptation alternatives (IPCC 2019). The Kyoto Protocol, an international agreement adopted in 1997 and effective in 2005, set binding targets for emissions reduction. Its first commitment period ran from 2008 to 2012, and its second commitment period from 2013 to 2020 (UN 2019a). The Paris Accord, adopted in 2016, aims to keep the global temperature in check; signatories commit to taking the necessary measures to reduce rise in temperature through several known practices such as reducing carbon emissions (UN 2019b).

There is a long history behind these most recent agreements focusing on climate change. In 1972, for the first time in history, the United Nations Conference on the Human Environment brought developed and developing countries together to discuss environmental issues that affect the entire planet, regardless of level of development and political boundaries (UN 1973). In the 1980s, Gro Harlem Brundtland, then Prime Minister of Norway, chaired the World Commission on Environment and Development (WCED), an independent commission created by the UN in 1983. The commission's report, "Our Common Future," also known as the Brundtland Report and published in 1987, addressed environmental issues and strategies leading to global sustainable development. It stated that "the environment does not exist as a sphere separate from human actions, ambitions, and needs, and therefore it should not be considered in isolation from human concerns. The environment is where we all live; and development is what we all do in attempting to improve our lot within that abode. The two are inseparable" (UN 1987, 7).

The first summit that brought together virtually every nation in the world to discuss and deal directly with sustainable development was the United Nations Conference on Environment and Development, aka Rio 92. The resolution document from that first summit was Agenda 21, which was adopted by 179 countries (UNCED 1993). Agenda 21 and the development of Local Agendas 21 in 113 countries initiated the process of transferring international resolutions from the international arena to the local level. Since then, the Rio+10 summit (World Summit on Sustainable

Development) took place in Johannesburg, South Africa, in 2002 and Rio+20 (United Nations Conference on Sustainable Development) took the summit back to Rio de Janeiro, Brazil, in 2012. The Rio+10 outcome document, Millennium Development Goals (MDGs), set targets to be met by 2015; the Rio+20 resolution was promulgated in 2015 in the form of Sustainable Development Goals (SDGs) with all targets expected to be reached by 2030.

Citizen Science Program: Science Helping Law Enforcement

Several Citizen Science projects have originated from communities' frustrations related to continuing environmental degradation. The lack of law enforcement is one of these triggers that bring communities together to achieve something that, ordinarily, would be up to governments and public authorities to address (McKinley et al. 2017). An example of this type of citizen-driven action is the Save Our Streams (SOS) program. Created in 1969 by a group of concerned sportsmen and conservationists interested in watershed stewardship, the group's charge was to monitor streams and rivers, raise public awareness of environmental abuses, and combat water pollution (Middleton 2001). Many other examples of volunteer monitoring and participatory environmental action, with varying levels of success, have been documented (Stepenuck and Green 2015). Benefits to participants and their communities usually include knowledge acquisition, change in attitudes and behavior, and more effective civic participation, including a voice in decision-making and influence on practices and policies.

A national survey conducted in 2013 collected data on volunteer monitoring programs that had been initiated between 1965 and 2012 (McKinley et al. 2017). A majority of respondents were involved in programs to effect change in environmental policy and to enforce regulations at both state and local levels. Examples include identification of illegal connections to municipal stormwater systems, improper wastewater discharges, and reporting to authorities of pollutant data in violation of the Clean Water Act. Citizen scientists may also partner with nongovernmental organizations (NGOs) to achieve their law enforcement objectives (fig. 17.3). One example of citizen scientists helping federal agencies is discussed in the next section.

FIGURE 17.3. Perth, Australia: citizen scientists monitor water quality and other indicators that help address environmental issues.

Expert Insight: Tonawanda, Western New York

Some of New York State's largest industrial facilities are located in Tonawanda, a suburb of Buffalo on the Niagara River. Similar to the case against the California utility company Pacific Gas and Electric Company (PG&E) portrayed in the 2000 movie *Erin Brockovich*, Tonawanda citizens linked chronic health problems in their community to environmental conditions. While the PG&E case was connected to water contamination, Tonawanda was experiencing air quality issues (McKinley et al. 2017). Air samples were collected by volunteers who organized themselves as the Clean Air Coalition of Western New York (CACWNY). The data showing elevated levels of a known carcinogen (benzene) were presented to state and federal regulatory agencies—the state's Department of Environmental Conservation (DEC) and the U.S. Environmental Protection Agency (EPA). DEC investigated further, and changes were made to improve air quality. In addition to its initial Tonawanda project, CACWNY is now involved in the Delavan Grider community and the Seneca Babcock neighborhood in Buffalo (CACWNY 2019).

18

Biophilia

Humans have an inherent attraction to nature. The biophilia hypothesis (Wilson 1984; Kellert and Wilson 1993; Kellert 2012) has been tested by several studies of built environments, and there is clear evidence that exposure to and contact with nature has numerous positive outcomes (Kaplan and Kaplan 1989; Ulrich et al. 1991; Chiesura 2004; Ward Thompson 2011). Research conducted in healthcare facilities, schools, and office buildings shows that patients heal faster, students perform at higher levels, and workers increase their productivity in buildings that integrate natural (biophilic) features in their indoor and outdoor spaces (Ulrich 1984; Heerwagen 2000; Kellert 2018). Physical and mental health benefits are strongly correlated with experiencing nature. Places with natural daylight and ventilation, and with windows that allow a visual connection with nature, have been shown to have a positive impact on health (Kaplan and Kaplan 1989; Heerwagen 2002). In fact, several studies conducted since the 1990s that simply show images or videos of nature to people in distress conclude that recovery time from trauma is reduced, stress and anger decrease, and heart rates and other physiological indicators stabilize (Ulrich et al. 1991; Kaplan 2001).

Despite all this evidence, most people living in urban areas today function in completely human-made environments. With an average of 80% of people in the developed world living in cities, many of us have no opportunity to come in contact with nature on a daily basis; our habitat is the built environment. Adults move between climate-controlled homes to climate-controlled offices and back. The elderly are confined to indoor environments. Children play in plastic playgrounds. In his book *Last Child in the Woods*, Richard Louv (2010) discusses the developmental benefits of exposing children to nature. He advocates for a return to old ways of playing, distancing children from electronic devices in favor of increased contact with plants and animals to instigate curiosity and encourage kids

FIGURE 18.1. Paris, France: people are inherently attracted to nature. Urban parks offer relaxation and respite.

to explore the natural world. More importantly, research shows that if humans are not exposed to natural environments when children, they will be less inclined to feel any affiliation with nature when adults (fig. 18.1).

What: Creating Biophilic Environments

Biophilia was first defined as "the innately emotional affiliation of human beings to other living organisms" (Wilson 1984, 31). Biophilic Design is defined as "biophilia applied to the design and development of the human built environment" Kellert (2018, 17). Designing human habitats that take this inherent attraction of human beings to nature into consideration results in built environments that improve not only the experience for people inhabiting them, but also their overall quality of life. In addition to different types of green infrastructure such as green roofs and green walls, several other design elements are used to create biophilic environments. Kellert (2018, 18–22) offers nine principles of biophilic design, arguing that biophilic environments are more than simply places with nature inserted into them, as follows:

- Biophilic design focuses on human adaptations to nature that advance physical and mental health, performance, and well-being.
- Biophilic design creates interrelated and integrated settings where the ecological whole is experienced more than its individual parts.
- Biophilic design encourages engagement and immersion in natural features and processes.
- Biophilic design is strengthened by satisfying a wide range of values that people inherently hold about the natural world.
- Successful biophilic design results in emotional attachments to structures, landscapes, and places.
- Biophilic design fosters feelings of membership in a community that includes both people and the non-human environment.
- Biophilic design occurs in a multiplicity of settings, including interior, exterior, and transitional spaces and landscapes.
- Effective biophilic design involves an "authentic" experience of nature, rather than one that is artificial or contrived.
- Biophilic design seeks to enhance the human relationship to natural systems and avoid adverse environmental impacts.

As with all nature, biophilic features can vary greatly depending on micro- and macro-climatic conditions. Cities like Singapore have become famous for their biophilic approach to development, in great part because the tropical climate of Singapore is highly favorable to lush vegetation, and the year-round level of precipitation is conducive to plant growth without much irrigation. Although most examples of biophilic design are of green elements, a tropical climate is not a necessary condition to create biophilic environments (fig. 18.2). In fact, some cities in Scandinavia and Nordic countries, where subarctic conditions predominate, have been lauded for their biophilic approach to development. One such example is the city of Malmö in Sweden, where the Green Roofs Research Centre devises new ways to make green roofs contribute to the quality of the environment, absorbing noxious pollutants and restoring health to highly contaminated areas (Beatley 2011). Also in Malmö, renewable energy produced at individual sites is not only contributing to a biophilic approach to development, but also transforming the dependency paradigm of the power supply.

A biophilic approach to urban structures improves quality of life in general and creates environments not only more desirable to live in, but

FIGURE 18.2. Vancouver, Canada: biophilic features, such as green walls, improve the aesthetic and functional quality of the University of British Columbia campus.

also conducive to restoration and rehabilitation. In fact, biophilic design strategies are being incorporated into several certification systems such as LEED (Kellert, Heerwagen, and Mador 2008). As our understanding of the ecological and social benefits of integrating natural and built environments evolves, we are better able to incorporate certain natural features into our buildings and cities. Traditional architecture and construction require excessive energy use; building systems that rely on nonrenewable energy contribute to cities being responsible for three-quarters of global greenhouse gas emissions (Elmqvist, Fragkias et al. 2013). Biophilic design is a sustainable design strategy and can mitigate the effect of urban structures on cities and allow humans who spend most of their lives in cities to have adequate exposure to and reconnect with natural environments.

So What: Benefits of Biophilic Design

Biophilic design enhances the experience of a place, and its positive outcomes have been recognized: "Biophilic design seeks to create good habitat for people as a biological organism in the built environment that advances

people's health, fitness and wellbeing" (Kellert and Calabrese 2015: 6). Geo-
metric patterns inspired by nature are commonly found in architecture,
regardless of style. Nature-based design was widely used long before the
benefits of biophilic design were identified. For example, architects like
Antonio Gaudí and Louis Sullivan have made extensive use of natural ana-
logues in the ornamentation of their buildings. Frank Lloyd Wright and
the Prairie School also adopted biomorphic forms in their designs (fig.
18.3). Le Corbusier's towers in the park were another example of attempts
to interconnect architecture with nature and provide opportunities to city
dwellers to have exposure to the natural environment despite living in an
urban, dense area. Some of these architects were reacting to the dehuman-
izing experience of industrial cities, but even modernists, who compared
buildings to machines, attempted to connect dwellers with nature by using

FIGURE 18.3.
Buffalo, New
York: bio-
morphic
elements in
the Darwin D.
Martin House,
designed by
Frank Lloyd
Wright and
built between
1904 and 1906.

natural materials and great spans of glass that visually connected the users with the outdoors. Vernacular architecture, although not explicitly based on biophilic principles, has also contributed to biophilic design by utilizing local materials and being responsive to climatic conditions of the site.

Biophilic design takes it farther than standard sustainability practices and is not limited to buildings and green infrastructure; it can be achieved in public infrastructure as well. Melbourne's Metro Tunnel Project, including five underground stations, is a recent example. A research project conducted by scholars at Deakin University investigated the possibilities for biophilia in these metro stations, and the Melbourne Metro Rail Authority decided to make this the first public transit project with biophilic features (Downton et al. 2017). In addition to the well-known benefits of public transit—reduced greenhouse gas emissions, accessible and widespread transportation networks, and improved mobility—adopting biophilic design for transit infrastructure helps improve the quality of urban environments even further. Beyond the immediate benefits from biophilic design, the Metro Tunnel Project is contributing to Melbourne's chances of becoming an "eco-city," an ambitious goal established in the Melbourne 2030 plan (State of Victoria 2002).

The benefits of biophilic design are not realized by simply introducing vegetation into buildings and their surroundings (Browning, Ryan, and Clancy 2014; Kellert and Calabrese 2015; Downton et al. 2017). In fact, the exposure to nature from which humans benefit may be achieved without any living organism. Research has demonstrated that indirect experiences of nature, such as images or scents, will trigger the same psychophysiological response in humans as direct contact with nature (Ulrich et al. 1991; Kaplan 2001). Additional cognitive benefits of biophilic design can be gleaned from a visual connection to nature. Nature in any space can be perceived by the senses, which also identify the feeling of respite and protection afforded by the sensation of prospect and refuge (Kellert, Heerwagen, and Mador 2008; Kellert and Calabrese 2015). Other physical elements, such as light, airflow, and temperatures, contribute to these sensations; more abstract elements include cultural connection to places and emotional relationships to natural features, patterns, motifs, and processes.

Now What: Making Biophilic Environments the Norm

Biophilic environments are more than a collection of green infrastructure, biomorphic elements, or natural materials. To achieve a biophilic environment, we need an ecological and integrated whole. There are several biophilic design attributes that, when aggregated and integrated into a space, will create the sensory experience of nature that results in well-being (Kellert and Calabrese 2015, 12–20): "light, air, water, plants, animals, weather, natural landscapes and ecosystems, fire, images of nature, natural materials, natural colors, naturalistic shapes and forms, natural geometries, prospect and refuge, organized complexity, integration of parts to wholes, and transitional spaces, among others." Biophilic environments result from the combination of biophilic elements at different scales integrated into an ecological whole; isolated design strategies are usually not successful in producing biophilic built environments.

Education plays a significant role in creating standards that become the norm. When children are exposed to environments that display the qualities desired to become the norm, they adapt and grow believing that what they do on a daily basis is the norm; it becomes part of their core values. Having schools that instill biophilic principles in children and young adults helps advance the environmental mission. One school that has applied this philosophy to its environments and pedagogy for almost 15 years is Sidwell Friends School (2019) in Washington, D.C. The school's belief is that students can better understand the impact of humans on the environment when they study in spaces that display how earth's resources are finite. The school believes in using "buildings as teachers": students learn how rainwater is recycled, electricity is generated, and energy conserved by simply experiencing the buildings they occupy and understanding what makes them "green" (Sidwell 2019). More proactive lessons are used in their curriculum; for example, science classes incorporate observation of rainwater moving through the green roof system. Sidwell Friends School has two LEED Platinum and three LEED Gold buildings (see chapter 12 for more on the U.S. Green Building Council). The school's LEED-certified buildings have green roofs and photovoltaic panels and use graywater. Other green building features comprise passive solar design, including natural ventilation and daylighting; the use of renewable materials, such as cork instead of linoleum on floors, and materials with recycled content and low chemical emissions; and the use of reclaimed wood for exterior cladding, interior wood paneling, windowsills, flooring and decking, gym bleachers, and barns (AIA 2019). Other biophilic elements were introduced, such as skylights added to existing buildings to increase exposure to natural light.

Sidwell's environmental ethic pervades not only the school, but also the extended community. In addition to environmentally friendly buildings

FIGURE 18.4. Vancouver, Canada: community gardens bring not only biophilic character, but also social and health benefits to a city.

and a constructed wetland for recycling and treating its wastewater, the school is easily accessible by bicycle and transit. Green Seal certified products are used to maintain the school's facilities, reducing its impact on the community at large (see chapter 12 for more on Green Seal). All materials used in the school, including electronic devices, are recycled. In addition to striving to reduce the impact of its physical facilities, Sidwell adopts other practices to make the school a biophilic environment. Students eat local in-season produce and hormone-free products and contribute their food scraps to compost piles used in their own gardens to fertilize the very food they eat (fig. 18.4). Only native species are planted throughout the grounds, minimizing the use of pesticides, reducing the need for irrigation, creating habitat for wildlife, and restoring native ecosystems in the periphery of the school. If more schools would adopt the practices implemented at Sidwell Friends School, our youth would become better stewards of the environment. Biophilic environments have a better chance to become the norm if future generations are brought up in them and learn from an early age that we can live in urban environments and still be part of nature.

Citizen Science Program: BioBlitz

Several cities in North America are monitoring the presence of nature in their urban areas with the help of citizen science programs. Citizens participate in BioBlitz groups to document plants and animals and determine the biodiversity of natural areas and urban parks. Initiated by the U.S. National Park Service in the 1990s (National Geographic n.d.), this movement has spread all over the world. In partnership with the National Geographic Society, thousands of observations were made using the iNaturalist app, now owned and managed in partnership with the California Academy of Sciences. Several other organizations, such as the Canadian Wildlife Federation (CWF 2021) and the Woodland Watch Project in Western Australia, encourage citizen scientists to get engaged in BioBlitz. In addition, urban areas are competing through City Nature Challenges (SciStarter 2017), also using the iNaturalist app, to capture the fauna and flora contributing to biodiversity in cities. The more biophilic our cities become, the more opportunities for nature to be integrated into the urban fabric and for wildlife to thrive in urban areas.

Case Study: Dockside Green

The Dockside Green neighborhood in Victoria, British Columbia, Canada, became famous as the first North American neighborhood to achieve LEED-ND (Leadership in Energy and Environmental Design—Neighborhood Development) Platinum certification. The development is located on Victoria's Inner Harbor and was a brownfield site used by light industry for more than a century. In addition to spilled petrochemicals, the site had been a landfill, and thus cleanup cost estimates were high. Vancity Credit Union, an organization known for innovative and socially conscious investment strategies, purchased the site from the city and initiated its development. The development began in 2005, but the global financial crisis of 2008–2009 stalled it. Dockside Green got new life in 2017 when the Victoria City Council voted to approve rezoning for the 15-acre site. A new master plan was created providing for a dozen new buildings, including seven residential towers. Following the rezoning, Vancouver-based Bosa Development bought Dockside Green from Vancity and committed to continuing to build the project to LEED-ND standards.

So far, Dockside Green has completed (1) 266 residential units, (2) three commercial buildings, (3) the on-site wastewater treatment plant and graywater recycling system, (4) a biomass district energy facility that generates heat for the entire development, (5) a greenway and waterway that connect the entire site, (6) a temporary space for interim uses such as urban agriculture, (7) art installations, and (8) public markets. The development adheres to the principles of sustainable architecture and green building and is expected to house more than 2,500 residents at full build-out. Transportation features that contribute to its sustainability include a car sharing program, linkages to a regional cycling trail, and bicycle racks and showers for commuters. Biophilic design elements include green roofs with public courtyard spaces, living walls, and filtration systems to purify and reuse water.

The original vision for the development has yet to be realized. The intention was to build a mixed-use, sustainable neighborhood that would contribute to the well-being of not only its own residents, but also residents in surrounding areas. The development, located in Victoria West, a traditionally low-income community, intended to act as a catalyst for a vibrant local economy. The affordable housing goal has not yet been met; only 49 of the 266 units built so far are social housing units, and, because many are one- and two-bedroom units, they are not appropriate for families with children. Despite its failure to provide affordable housing, Dockside Green has succeeded in rehabilitating a contaminated site, building a wastewater treatment plant and biomass energy production facility, and creating a mixed, walkable, and connected neighborhood (Webb 2009; Descoteau 2017; Duffy 2017).

19

Attachment and Place-making

Urban green spaces embody a variety of meanings for people depending on the relationship a person has with the landscape. A park used for daily walking might represent health and well-being; other green spaces may have spiritual meaning based on cultural or historic significance; and others, such as private yards or neighborhoods, may represent the safety and security of home and a feeling of pride of ownership and responsibility toward nature. It is the emotional bond with nature that creates these meanings. Common heritage and cultural values with shared meaning are often reflected in our urban green spaces that contribute to the identity of the city. Some cities are known for their topography and views, others are known for their bays, rivers, and beaches. Some are famous for their human-made landscapes; Central Park in New York City and the green squares in Savannah, Georgia are examples of green spaces that have deep meaning and identity for individuals and their cities. Urban nature includes both natural areas and built green spaces such as streetscapes (fig. 19.1), green roofs, and green walls. Factors that make green spaces preferred and meaningful include restorative effects, recreation opportunities, familiarity, emotional importance, and community significance (Wolf, Krueger, and Flora 2014, 2–3).

Meaning and identity link an individual with the physical environment, a significant concept for urban ecology. Planners and designers should understand how people form bonds with nature to help them create meaningful green spaces. Place attachment, created by person-to-place emotional bonds, is called topophilia, which refers to an individual's love for certain aspects of a place (Wolf, Krueger, and Flora 2014, 2, 5). Two components of a place generate topophilia: the physical attributes, both built and natural, and the social attributes that facilitate social relationships. Place attachment can make people feel more responsible for the environment, especially if a meaningful natural environment is associated with personal

FIGURE 19.1. Livermore, California: community identity with unique vine trellis for streetscape. The City of Livermore Specific Plan (form-based code) was established to create a new identity for the downtown district.

benefit, which can lead to pro-environment stewardship activities (Wolf, Krueger, and Flora 2014, 7). People who have positive emotions about a place tend to actively protect and care for the place, which is why the homogenization of urban landscapes, which creates places with no meaning, is a concern for the future of sustainability in urban areas (Kaymaz 2013).

Experts in ethnobotany, archaeology, and ethnoecology can help discover meaning and identity by researching indigenous cultures and historic land cover to help guide the process of place-making; however, research, facts, and statistics can only go so far in making people care about places and place-making. People, with the exception perhaps of some scientists, typically do not think about plants, animals, and landscapes in terms of statistics; they relate to landscapes on an emotional level, and their attitudes toward nature inspire them to act responsibly toward the environment. This means we also need experts, such as science writers, journalists, social anthropologists, and environmental psychologists and philosophers, who can write an emotionally compelling story of science that will make people care about the environment and motivate positive behavior. For example, one way to tell a story is to observe ways that traditional people developed strategies to sustain environmental services and convince people to care about nature (Hummel 2016). Understanding meaning and sense-of-place in cities is the perfect setting for citizen science projects. Data gathering can include personal observations by participants that give more meaning to the process and outcome of the project. For example, a tree ID project in Melbourne, Australia that encouraged people to report tree problems had the surprising outcome of people writing love letters and encouragement to trees with which they had created an emotional bond as a result of interaction with the tree (LaFrance 2015). The process of place-making is a citizen-based participatory planning and design process that could incorporate citizen science data gathering.

What: Landscape Meaning and Place-making

Green spaces are symbolic of a time and place and reflect cultural and social norms, including attitudes and behavior toward nature. Neighborhood landscapes in particular are personal expressions that change with time and owners but often retain remnants of legacy landscapes—the historical landscapes that came before them. Nature in landscapes is always evolving and changing, which provides opportunities for re-creating, improving, conserving, and protecting public spaces in a process called place-making. It is important to distinguish between space and place, two terms often used interchangeably in urban areas; a *space*, such as a green space, has spatial location with dimensions and objects, but it becomes a *place* when society and individuals assign meaning and value to it. The concept of place-making is important for urban ecology; it presents a significant opportunity to strengthen community identity through the natural features of a place. While place-making is a way to improve a city by reimagining and reinventing public spaces in cities (fig 19.2), it pays particular attention to the physical, cultural, and social identities that have defined a city historically, which puts ecology at the center of the place-making process.

Place-making depends on a collective vision of the people who use the space and the concept of contributing to the well-being of citizens. Project for Public Spaces (PPS) is an organization that promotes place-making and provides services to communities to help them with the process of creating a vision for their community. Their collaborative community process is centered on observing the people who use the space, listening to them, and asking questions to understand their needs. PPS designated four key categories to illustrate the characteristics that make a place great: Sociability, Uses and Activities, Comfort and Image, and Access and Linkages. Two categories, Uses and Activities and Comfort and Image,

FIGURE 19.2. Livermore, California: a pocket park along a newly renovated main street enhances city function and identity.

are particularly important for urban landscapes and ecology. Key terms in each category describe the intangible qualities of a space that can guide the design and function of green spaces. For example, the qualities that fall under Uses and Activities include active, special, real, useful, indigenous, celebratory, fun, vital, and sustainable. All of these qualities can be applied to parks and green spaces in the design process as a way to ensure the design is meeting ecological and social goals for the community. Plant material, in particular, can make a place feel special, real, and vital. The use of native plants can honor the indigenous quality of the space, celebrating its cultural and historic heritage. Plants can also create useful spaces that are fun, vital, and active while achieving sustainability goals for the city. Qualities in the Comfort and Image category, such as safe, "green," spiritual, charming, attractive, historic (fig. 19.3), and walkable, also support best design practices for green spaces. The list describes everything a community park should be, including activities that will make it functional and special. To encourage people to use green spaces and develop a stronger ecological identity, the spaces must be convenient, accessible, and within close proximity, three qualities in the Access and Linkage category. Qualities found in the category of Sociability are important because they speak to our emotions and feelings that encourage the use of green spaces, such

FIGURE 19.3. Vienna, Austria: historic Hundertwasserhaus, famous for rooftop gardens and greenery, is part of Austria's cultural heritage.

as welcoming, interactive, friendly, and neighborly. Other qualities such as stewardship, cooperation, and pride encourage active volunteerism and participation in community projects (Project for Public Spaces 2007). Preserving natural and cultural resources and the environment is an important principle in place-making. Capturing the unique sense of place means landscapes should grow from the local climate and topography (Grabow 2015, 17, 25).

So What: The Impact of Place-making on Urban Ecology

Place-making can contribute to ecological and environmental justice, by making people more aware of their surroundings and how their activities contribute to environmental health. People form their own ecological identities through different activities that help them understand what it means to participate as an urban citizen in improving the environment. Activities include storytelling, interpretation, and representing places through photography, art, and music, as well as participating in community gardening or restoration of natural areas that can promote community engagement. Our ecological identity, one of many human identities, reflects the perspective we have on ecology and our own personal responsibility for sustainability and the environment. An ecological identity (fig. 19.4) that focuses on the environment in cities can help people feel empowered to create better green spaces to improve biodiversity and ecosystem services (Adams et al. 2016).

Place-making is important to urban ecology because people behave differently toward landscapes they care about. Landscapes with special meaning are better maintained and protected from harm. An important concept in place-making is to recognize the landscape characteristics that play a role in people's perceptions and preference for landscapes. For example, research has shown that people have a preference for and emotional response to trees with wide canopies. One theory is that trees with this shape on the African savannas, where humans evolved, were an important cue to the location of water necessary for human habitation, establishing an evolutionary preference for wide-canopied trees. Studies have also shown that people have preferences for and positive emotional responses to different tree colors, with green and red trees being preferred to purple or brown trees. The theory of learned evolutionary knowledge includes leaf color as a cue to the nutritional value of the plant, a behavior documented in other primates who select primarily green and red leaves with high nutritional value (Kaufman and Lohr 2004). The practical aspect of this information is that selection of plants that elicit a positive emotional response to landscapes is a way to change people's perceptions and behavior toward a landscape. The key is to match emotionally preferred plants or features with plants of high ecological value to create a landscape that will contribute to ecology and preferred places.

FIGURE 19.4. Perth, Australia: creating ecological identity with native drought-tolerant plants on Curtin University campus.

Now What: A New Role for Meaning and Emotion in Urban Ecology

The relationship between emotions, place, health, and urban green spaces is gaining more attention in public health policy and urban design. The focus is on mental and emotional health benefits of green spaces for children and adults. Studies have shown that nature experience is associated with psychological well-being, including a positive effect on happiness, social interactions, and mental stress. Nature can also give a sense of meaning and purpose to life and improve cognitive function such as memory and attention. Children often show improvements in imagination and creativity with exposure to nature, which improves school performance and impulse inhibition (Bratman et al. 2019, 3). With no opportunities for exposure to nature, some are concerned that children will never develop attachment or love for nature, and that lack will prevent an entire generation from developing pro-environmental behaviors. Middle childhood tends to be an important time for place attachment, because feelings of connection at an early age tend to grow stronger in later years. Children often prefer green, natural settings, and they form attachment and meaning through activities within a place (Wolf, Krueger, and Flora 2014, 7). The types of activities and experience with nature are important; fun and gratifying exposure to nature increases emotional connections, but there is also evidence that urban children can find nature scary and uncomfortable, with fears of getting lost or attacked by insects or animals. Children who have a positive view about spending more time in nature score higher in eco-awareness and pro-environmental behaviors. Research has also shown that the more time spent in nature as a child, the more positive the influence on environmental attitudes and the more pro-environmental a person is during childhood and as an adult (Collado et al. 2015, 66).

The concern about children lacking pro-environmental attitudes is their ability to be good stewards of the land for future sustainable and resilient cities. The new frontier in planning is to use biophilic design strategies to create greener cities where people are near a green space or park at any given time and spend more time in outside activities. Biophilic design uses nature to create spaces where positive emotional experiences, such as enjoyment, interest, pleasure, and wonder generate attachment and caring for a place (see chapter 18 for further discussion about biophilia) (Wolf, Krueger, and Flora 2014, 8).

Citizen Science Program: Shmapped Project

The Improving Wellbeing Through Urban Nature (IWUN) project at the University of Sheffield developed a smartphone app, called Shmapped, to show how urban environments affect people's health and well-being. The app, launched in July 2017, prompts city residents to notice and map the good things they encounter in their city by recording the user's locations and activities. The hope is the maps can reveal new insights into the ways different spaces affect residents' well-being. The goal of the project is to help designers, planners, and public health organizations create better urban spaces for people and wildlife. The app allows people to instantly respond to their environment and helps researchers by identifying the precise location and the characteristics of the place with notes and photographs. The app also collects baseline well-being data with a brief questionnaire. Knowing what it is about spaces that benefit well-being will help in creation of better urban spaces for people and wildlife (Barton 2017).

Projects: MINDSPACES

The MINDSPACES project in the European Union is using virtual reality and scenario-testing models to help urban planners get real-time feedback about the impact of their designs on mental health by testing people's emotional response to spaces such as urban plazas and redeveloped buildings before the projects are built. The idea is to design a city that can boost well-being. A team of neurologists, architects, artists, and epidemiologists develop designs for spaces such as city squares which they turn into a digital simulation that locals view using virtual reality (VR) goggles. While exploring the virtual space, users also wear devices that measure their brain activity, skin response, and heart rate. Neurologists use the data with machine-learning programs to figure out the most pleasant, inspiring, or emotionally appealing aspects of the proposed design. Based on participants' emotions, the space is redesigned in real time; for example, items they don't like are removed from the scene and other items inserted until the design is judged "emotionally friendly." The project is currently aimed at the spaces used by elderly people, and one of the test cases involved redesigning a home for seniors in Paris, France (Burke 2019).

20

Landscape Behaviors

Landscape behaviors are those actions and activities humans engage in that have an impact on the environment, including how people create and use the landscape, how they care for it, and how they advocate for the protection and preservation of the landscape and all it supports. Social institutions (those rules that govern peoples' actions) often drive landscape behaviors, including informal institutions such as unwritten norms, customs, and expectations, and formal institutions such as policies, regulations, and laws (Larson and Brumand 2014). The norms and customs of informal institutions include those landscape behaviors that reflect personal values and beliefs, such as environmental ethics, aesthetic preferences, property ownership, social responsibility, personal health concerns, and socioeconomic status. The formal institutions of rules and laws are mandated behaviors and activities that reflect common societal values for health, safety, and welfare interests, such as irrigation restrictions, fertilizer bans, and ordinances in neighborhood communities (Sisser et al. 2016). The study of landscape behaviors is a subset of the study of environmental behavior. Theories and models from sociology, psychology, and anthropology are used in the context of ecology, conservation, urban studies, and social welfare to create an emerging field of study called environmental and conservation psychology, with a focus on socioenvironmental behaviors.

Behaviors can have both positive and negative effects. Recognizing and changing negative behaviors can be difficult even when people are presented with compelling evidence for the need for new behaviors. Changing behaviors requires recognizing barriers, acquiring new knowledge, and identifying relevant incentives and do-able actions. Environmental education was once considered the key to pro-environmental behavior because of its capacity to change awareness and knowledge; however, studies have shown that emotions—what people feel and believe about the environment—rather than knowledge, determine their attitude and the likelihood of demonstrating pro-environmental behavior (Pooley and O'Conner 2000). Several barriers to behavior change have been identified. Internal barriers include a persons' attitude or environmental concern, such as lack of environmental consciousness, a lack of internal incentives, and the feeling that one cannot influence the situation or should not have to take responsibility for it. External barriers include the influence of social, cultural, or family norms, political or infrastructure barriers, and constraints such as lack of time, money, or information (Kollmuss and Agyeman 2010). New approaches to behavior change include the use of social marketing principles to encourage new behavior and use of verbal or written commitments to take action. The key is to identify the issues that resonate and people's reservations or inclinations toward a behavior to develop strategies, such as educational programs and social marketing messages, to encourage behavior change (Warner and Monaghan 2016).

Citizen science projects and volunteer programs typically engage people whose values and beliefs include a pro-environmental ethic, and an individual sense of personal responsibility for pro-environmental behavior. CS projects provide opportunities for ordinary people to engage in large-scale data collection that can contribute to meaningful environmental programs, protections, and policies. The advantage of CS programs is that they often use those elements known to motivate people to formulate projects and encourage participation. For example, people are more likely to act if they are aware of the issues and if they know what actions to take. They are also more willing to act if they perceive they have the ability and personal responsibility to bring about change, and they make a written or verbal commitment to take action (Kollmuss and Agyeman 2010).

What: Motivations for Pro-Environmental Landscape Behaviors

Pro-environmental behaviors are usually described as individual behaviors that consciously seek to minimize the negative impact of personal actions on the environment, such as minimizing energy and resource consumption, reducing waste, and reducing use of toxic materials (Kollmuss and Agyeman 2010). However, behaviors can also include positive impacts such as protecting or creating natural habitat, advocating for protection of species and pro-environmental policies, and involvement in political and community environmental activities.

Several models and theories have been proposed to explain pro-environmental behaviors; most are based on psychological or sociological theories about general human behavior, but because of the complexity of the human/environment relationship, no single model has emerged to explain pro-environmental behavior. Early linear models, which have been proven wrong, proposed that environmental knowledge would lead to awareness and pro-environmental behavior. Other models identified factors for environmental behavior such as attitudes, personality, sense of personal responsibility, and possession of action skills, yet they failed to explain how the relationship between the factors would lead to certain behaviors. Social models of altruism (compassion for the well-being of others), empathy, and pro-social behavior have also been used to explain pro-environmental behavior. Some suggest altruism is needed to focus beyond oneself and care for the wider community. Others have proposed that three orientations: altruistic (removal of harm to others), egoistic (removal of harm to oneself), and biospheric (removal of harm to the non-human world) are needed for pro-environmental behavior. Other sociological models propose a variety of social and psychological factors to explain behavior, such as environmental values and awareness, behavioral incentives, and external opportunities to act ecologically. Ability to understand consequences of behavior and access to knowledge that modifies attitudes and values may also explain behavior. Social scientists emphasize that many pro-environmental models are limited because they fail to take into account constraints, and they often erroneously assume humans are rational—meaning that appeals by scientists and environmentalists to change behaviors can often be largely ineffective (Dorsey 2010; Kollmuss and Agyeman 2010).

It is generally agreed that a useful model is nearly impossible to create from the numerous theories. Most researchers rely on a set of identified factors, including economic, social, and demographic, to help understand motivations and create targeted education and social marketing programs. External factors include infrastructure and codes that encourage pro-environmental behavior or prevent harmful behaviors, such as landscape codes that support the use of native plants. Economic factors include personal incomes and municipal policies such as rebate and tax incentives (or fines) for select activities. Social and cultural factors include neighborhood norms that encourage mimicry (replication) of yard designs (fig. 20.1), and cultural values that inform large-scale environmental policies. Demographic factors include income, education level, and gender. For example, studies show that women tend to be more emotionally engaged, environmentally concerned, and willing to change environmental

FIGURE 20.1. Hercules, California: mimicry in front yards reflects social factors, such as landscape codes and neighborhood norms, that influence residential design.

behaviors. Internal factors include motivation, priorities, values, attitudes, preferences, awareness, and emotions. When responsibilities and priorities, such as our own well-being and our family's well-being, align with pro-environmental behaviors, people are more motivated to act; however, if priorities don't align, motivation is low. Personal values and attitudes are tied to childhood experiences with nature, family values, and role models that help shape an environmental ethic. For example, the preference for green lawns is usually tied to the familiar—people often prefer green lawns because of childhood memories and aesthetic appeal. Environmental awareness is an understanding of the impacts of human actions on the environment, through either knowledge or perception. However, awareness may have little influence on behavior, because the complexity and slow pace of environmental change often leads to underestimating the consequences and reacting only to drastic and sudden changes. Emotional involvement is the ability to care about, and react to, environmental problems; however, emotional reactions can also lead to feelings of helplessness and certain defense mechanisms such as denial, apathy, and resignation that prevent pro-environmental behavior (Chowdhury et al. 2011; Dorsey 2010; Kollmuss and Agyeman 2010).

So What: Impact of Landscape Behaviors on the Environment

Landscape behaviors can be displayed in different ways by the average urban homeowner or landscape professional. These ways might include choices of plant materials and design and maintenance of the landscape (fig. 20.2). The most common environmental issues impacted by landscape behaviors are (1) wildlife loss due to chemical use; (2) poor water quality due to overuse of fertilizer; (3) lack of biodiversity due to poor plant choices, (4) overuse of plants that require high water input; and (5) poor horticultural practices that affect the health of plants.

Social norms, such as the desire to fit in with the aesthetics of the neighborhood, often influence poor plant selection and maintenance practices. The perception that people are judged by the appearance of their yards often compels people to purchase plants for aesthetics rather than for ecological function. It also creates a need to maintain landscapes at a neighborhood standard that often has unintended consequences for environmental health (Mooney 2015). Wealth is also a social factor that can influence decisions on the personal and neighborhood level. For example, wealth, or lack of it, can create an "income ecology" or "status ecology" based on the type and number of plants found in high- and low-income neighborhoods. Contrary to popular belief, studies have shown that low-income landscapes are often ecologically healthier due to fewer maintenance inputs, and high-income yards tend to be less healthy due to more inputs to maintain strict aesthetic standards. Higher expectations for perfection in these yards often lead to this condition, despite the fact that higher income homeowners tend to be more educated and often have more knowledge of the consequences of their actions. The pressure to mimic or conform to certain visual landscape qualities is also greater in high-income neighborhoods, despite the ability to afford a greater variety or number of plants that would increase biodiversity (Dorsey 2010).

Priorities such as safety and security concerns also influence behaviors; well-maintained landscapes are a sign that people are present and care about their neighborhood. Safety concerns often motivate people to use crime prevention techniques such as low shrubs, landscape lighting, and open turf so they can see and be seen. Another safety issue is fear about certain wildlife. People often desire open turf to see animals such as snakes, and they use pesticides to deter unwanted wildlife, particularly insects such as mosquitoes, often with negative consequences for beneficial wildlife.

FIGURE 20.2. Watercolor, Florida: a residential yard with no turf and native plants influenced by neighborhood plant material.

Ordinances and policies are external factors that influence decisions at three levels: individual property maintenance decisions, neighborhood-level HOA decisions, and municipal-level management decisions that inform policies (Sisser et al. 2016). Behaviors are not always by choice; HOA landscape codes often dictate plant choices for a cohesive aesthetic look throughout the community, which can reduce biodiversity. Many HOAs are also "maintenance free," a term that means landscape maintenance costs are included in the HOA fees and the same company maintains all landscapes in the community. This practice can create environmental problems if the company policy is to use fertilizers or pesticides on a regular schedule rather than as needed, or if the company spreads diseases and pathogens around the neighborhood by failing to clean equipment before maintaining each yard.

Confusion about proper maintenance activities, sometimes due to lack of knowledge, but also due to conflicting research or advice from different sources, can influence behaviors. While new technologies, such as smart irrigation controllers, promise water savings, many homeowners leave the controls as they were set by the home builder, rather than adjusting for seasons or restrictions. Failing to adjust controllers reflects an economic-, knowledge-, and priority-based behavior that wastes water. Advertising and cost of products can also influence consumer behavior; for example, weed and feed products, sold as time-savers in the garden, are not the best choice where more targeted approaches would be more effective (Dorsey 2010). Landscape choices can also be influenced by popular media that promotes ecologically poor design trends. Other common sources on which people rely for information include relatives or neighbors they trust or retail workers who sell landscape products (Blaine et al. 2012).

Lack of awareness of policies promoting positive behaviors, such as restrictions on lawn irrigation, are a leading cause of water quality and quantity issues. Most studies show that typically one-third to nearly half of homeowners are not aware of policies, and those who are often admit to not following the ordinance. Many don't understand the rationale or are not concerned about enforcement, noting that lack of enforcement means a small probability of being fined. The biggest challenge to code enforcement for most cities is lack of resources, including money and personnel to promote and enforce codes (Sisser et al. 2016).

Some homeowners can experience conflicts when making decisions between long-standing neighborhood norms and more ecologically responsible landscapes. For example, the norm of neatly mown, weed-free green lawns (fig. 20.3) often uses high amounts of water, fertilizer, and pesticides at an estimated mitigation cost of US$2.2 billion a year for U.S. water supplies (Mooney 2015). Yet, for many homeowners a lawn is a financial and social investment, and they hesitate to change because a naturalized lawn (turf with other plants/weeds that can be mowed) often goes against neighborhood aesthetic norms. They also seek approval of neighbors who consider a green lawn a reflection of good character, civic responsibility, and personal values. Many homeowners also apply chemicals to lawns in spite of knowing the negative impacts and having pro-environmental values because their materialistic and status attitudes toward their property and neighbors are stronger than their environmental attitude (Dorsey 2010; Sisser et al. 2016). Although many municipalities and neighborhoods recognize the value of naturalized lawns, they also hesitate to go against the norms for several reasons, including the challenging process of updating ordinance language that describes natural lawns and enforcing new codes (especially for code personnel with no horticultural knowledge), and the fear of creating loopholes that lead to poorly managed properties (Dorsey 2010; Sisser et al. 2016).

FIGURE 20.3. Sarasota, Florida: design of typical residential front yards with a continuous strip of weed-free green lawns.

Now What: Strategies for Landscape Behavior Change

Several strategies are being employed by communities, government agencies, university extension programs, and environmental organizations to increase pro-environmental behaviors. Targeted behaviors include use of landscapes that require high consumption of water, the high use of chemicals, and landscapes that diminish urban wildlife habitat. The primary strategies include using social media and social marketing campaigns, changing environmental policies and regulations, offering economic incentives, and engaging ordinary citizens in the research and scientific process through citizen science.

Scientists have stepped up efforts to learn more about environmental behaviors to help develop and inform strategies. Long-standing techniques such as interviews and focus groups are the most popular methods used to collect data, but new technologies such as data transcription software, eye-tracking devices, internet chat rooms, and activity-tracking devices are being used to collect additional data. Citizen science projects are also being used to collect data, create social marketing messages, and engage people in the study of their own behaviors.

The most promising strategies include social marketing, economic incentives, and legal ordinances. Social marketing uses principles of commercial marketing to develop programs aimed at encouraging positive behavior change to benefit human well-being. Environmental messages focus on the benefits to people (why they should engage in a behavior) and the benefits to the environment. To be successful the process must first identify three things: the barriers to change, the desired behavior change, and the audience's perception of that behavior (Warner, Ali, and Chaudhary 2017; Warner and Monaghan 2016). Messages should address commonly cited barriers and cite relevant, do-able actions to overcome them. University extension programs frequently use social marketing with clients, and many have active social marketing research projects. For example, a recent study revealed that residents value their landscape primarily for aesthetics and well-being benefits, but those who value aesthetics are least likely to engage in irrigation best practices, while those who value their landscape for habitat and well-being are most likely to engage in irrigation best practices. This information can be used to create a message that emphasizes how conserving water is compatible with a visually attractive yard that also serves habitat and health needs. One strategy used in social marketing is commitment, such as a pledge or promise to take some action. Commitments are usually public declarations such as yard signs or bumper stickers that reinforce the responsibility people feel to follow through on their intent (Warner, Ali, and Chaudhary 2017; Warner and Monaghan 2016).

Cities are increasingly adopting economic incentives (rebates) or disincentives (fines) to change behaviors. Programs in twelve cities in the U.S. desert southwest (Arizona) offer tax credits or rebates between US$50 and $3,000 to residents who replace turf with native and naturalized drought-tolerant plants. Others offer rebates between US$75 and $2,000 to change irrigation systems, including converting to graywater irrigation, installing irrigation timers, rain sensors, or smart irrigation controllers, or replacing irrigation heads with high-efficiency nozzles or drip irrigation. One city will pay up to US$1,500 to remove swimming pools or spas (Arizona Water Awareness n.d.).

Landscape ordinances and codes that mandate certain behaviors are increasingly using science-based and design-based language to set policy. One example is the Tampa Bay, Florida, city development codes. A chapter in the codes addresses sustainable environmental elements, noting that a healthy, sustainable environment is fundamental to creating a livable and prosperous city, and poorly managed growth can lead to loss of vegetation and ecological areas. Strategies include water-saving measures, investments to improve water quality, preservation of natural spaces for wildlife habitat, conservation of natural resources, developing pro-environmental attitudes, and balancing economic costs and benefits. A few examples of many behaviors that are mandated include the use of 60% native plant material in new landscapes, preservation of existing native vegetation, the provision of wildlife corridors, and preservation and protection of major wetlands in the city with a 25-foot buffer from development (Hillsborough County 2016).

Citizen Science Program: Public Participation in Scientific Research

Public participation in scientific research (PPSR or CSPs, citizen science projects) in natural areas has been shown to help improve knowledge and understanding of ecology and the environment. Recent research shows that PPSR experiences influence, and are influenced by, a "sense of place" that has the potential to change individual perspectives, attitudes, and behaviors about the environment and human/environment interactions (Haywood 2014). This finding implies that participation in citizen science programs could be another strategy, in addition to social marketing and economic incentives, to encourage pro-environmental behaviors

Expert Insight: Barriers to Changing Landscape Behavior

At the Fifth National Conference for Non-point Source and Stormwater Outreach (Portland, Oregon, 2009) several researchers and industry professionals convened a panel discussion on "Barriers to Changing Landscape Behaviors in Residential Settings." Behaviors discussed included irrigation practices on lawns, the use of chemical herbicides and pesticides in landscapes, the use of lawn fertilizers, and substituting native/resource-efficient (drought tolerant, pest free, hardy) plants for turf. Topics included attitudes and social norms, motivating beliefs (incentives), discouraging beliefs (disincentives), and situations that impact control, such as HOA codes and knowledge, costs, and ability to implement. Researchers shared insights they gained from homeowner interviews and focus group studies with owners who were primarily responsible for their own yard maintenance. On the subject of changing their landscape design to decrease turf and increase use of native/resource-efficient plants, panel members shared the following comments regarding perceived barriers from homeowners: (1) native plants look unkempt and neglected, presenting security issues (looks like no one lives here), (2) native plants look weedy and messy, not attractive, (3) homes will lack curb appeal and be hard to sell, (4) unwanted wildlife, such as snakes and rats, will come with native plants, (5) maintenance will cost more or won't know how to care for the plants, (6) don't know how to design the yard or prepare the beds for native plants, (7) don't know how to find adequate plant material, (8) the HOA won't permit the use of native plants or requires an approval process, (9) concerned they will hate it when it's done, (10) don't understand why native plants are so important to ecology or biodiversity, and (11) various concerns about social norms and economic status such as fitting in with the neighborhood and being good neighbors. All of these comments reflect the identified factors for barriers to pro-environmental behavior and provide insight that educators and social marketing programs can use to develop behavior change messages (author's personal conference panel notes).

21

Cultural Beliefs and Ethics

The role of environmental ethics in urban ecology is clear: ethics provide the moral grounds for social and environmental policies for more sustainable, environmentally healthy cities. Science-based information is needed for effective policies; however, science alone doesn't inspire or compel us to care. Science provides us with important information and knowledge, but we need ethics to strengthen science by considering the social implications of human values, morals, and beliefs for decision-making (Warner and DeCosse 2009a). While we know that ethics are important to guide the use of science, less is known about how a person's ethics influence their decision-making and how individuals acquire their environmental principles. Developing personal environmental morals and ethics is often an iterative process of learning and understanding through repeated exposure to concepts and ideas, making it difficult to know exactly which conditions influence ethics. Factors that can influence environmental ethics include:

- family beliefs and morals;
- exposure to nature (especially at a young age);
- personal values, and cultural and religious beliefs;
- emotions, education, and knowledge;
- community morals and ethics; and
- attitudes about personal responsibility and personal ability (fig. 21.1) to do something meaningful to help the environment.

Perhaps the simplest and most profound explanation of an ethical relationship with nature was written by Aldo Leopold, a prominent American environmentalist, in *Sand County Almanac* (1949), where he states, "we can be ethical only in relation to something that we can see, feel, understand, love or otherwise have faith in" (Warner and DeCrosse 2009b). Humans value things they love and things that are important to them, and they use

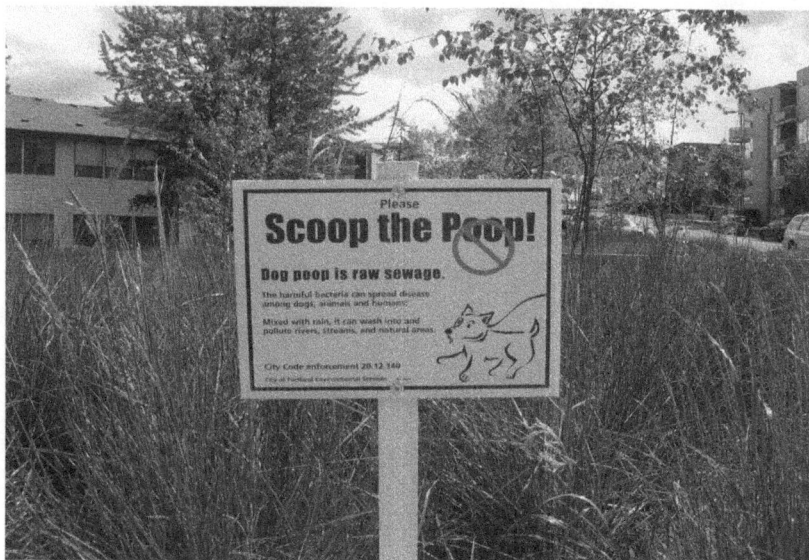

FIGURE 21.1. Portland, Oregon: a sign encouraging people to pick up after their pets is an example of encouraging ethical environmental behavior by appealing to the perceived value people place on the environment.

the perceived or assigned value of these things to make decisions about whether something is right or wrong, forming a basis for ethical behavior.

Which brings up a primary debate in environmental ethics: how should nature be valued? Do we value nature for the provisions that promote human well-being, or simply because nature exists and has rights? Cultural beliefs, those tenets and convictions that people in a community hold to be true, are often central to this debate, because cultural beliefs are the basis for the concept that nature and everything in it has rights. The idea that nature has rights derived from the cultural beliefs of indigenous peoples who believed that humans and nature are equal and interdependent, and humans are simply custodians of the land. The rights of nature and ecological governance are unique to different cultures that have a tradition of conservation and try to live sustainably within nature's limits (La Follette 2019; UN Environment Programme 2017). However, cultural beliefs must be supported by personal awareness and environmental ethics that come from emotional bonds with nature that create meaning, value, and place attachment for individuals (fig. 21.2). In this way environmental ethics are significant for urban ecology. A sense of right or wrong that comes from emotional connections and cultural beliefs is the force that compels people

FIGURE 21.2. Harbin, China: cultural beliefs are often manifested in nature when humans trim plants, such as trees, to create meaning through cultural aesthetic values.

to be involved in community activities and governance, especially when nature and natural areas are threatened or lost to development.

Experts who study the influence of ethics and cultural beliefs on our environmental decisions and behavior include those in the social sciences such as philosophers, sociologists, and psychologists. Others closely aligned with the social sciences include theologians, clergy, and religious officials who study rights of nature as a faith-based concept. Politicians and those engaged in environmental law are also involved in ethical debates. Others who have spiritual and cultural interests in land use include aboriginal people and environmental activists. Environmental ethics go to the very core of the concept of citizen science, where people who participate are usually driven by a social conscience. Those who engage have, at the very least, an interest in the environment, and most feel compelled to contribute because of a personal belief in the ethical and moral responsibilities of humans to protect nature and the environment.

What: Environmental Ethics and Cultural Beliefs

Environmental ethics is the discipline in philosophy that studies the moral relationship of human beings to nature and non-humans in the environment. The practical application is the ability of humans to recognize plant and animal rights and realize the intrinsic worth of nature. The practical purpose is to provide moral grounds for social policies aimed at protecting the environment and improving environmental degradation (Brennan 2015). Three key concepts of environmental ethics should be considered: the earth and its creatures have moral status; the earth and its creatures have intrinsic value; and humans should consider an ecosystem perspective that includes other forms of life in ethical thinking (Warner and De-Crosse 2009a). It is important to distinguish between *instrumental value* and *intrinsic value* as they relate to beliefs and values. Instrumental value is the value of something as a means to an end. For example, a wild plant may have value because it provides ingredients for medicine or is aesthetically beautiful; on the other hand, another wild plant, with no measurable provisions or benefits, may simply have intrinsic value because it exists, as an end in itself. When something possesses intrinsic value, it generates prima facie (accepted as correct) moral duty on the part of humans (moral agents) to protect it or refrain from damaging it (Brennan 2015). When thinking about the environment, some people assign intrinsic value to humans alone, an assumed moral superiority of humans (strongly anthropocentric), or they assign greater intrinsic value to humans than to non-human things, valuing them only if they promote human well-being or human interests (weakly anthropocentric).

How people think through a decision based on environmental ethics is important. Peoples' values and beliefs can cause them to ignore, reject, or distort scientific evidence if it doesn't support their goals or agenda. At the

FIGURE 21.3. Perth, Australia: cisterns at the Perth Zoo are an example of ethical environmental responsibility to manage and mitigate the effects of drought by collecting and storing water.

same time, people who value science and have a deep belief in the ability of science to solve problems may also ignore or distort the reality of some situations that need a more nuanced approach when human-centered factors impact the environment. For example, if we believe all participants have moral status, it forces us to consider the range of stakeholders now and in the future. It also requires consideration of the effect on stakeholders near and far, including humans, animals, and plants. The "when" requires considering cumulative long-lasting impacts and immediate impacts. Other factors that inform our reasoning are the notion of intrinsic value and the roles of risk, uncertainty, probability, and prediction. Reasoning about environmental ethics in decision-making often requires considering the impact of cumulative actions far into the future, but the accuracy of predictions can be questionable; ecosystems and nature are complicated, making it hard to determine the probability of predicted effects on the environment actually happening. The quality of the evidence used to assess risks and make predictions is important; if evidence is reliable and there exists a high probability of a damaging effect, the more powerful the ethical responsibility to manage or mitigate the effect (fig. 21.3) (Warner and DeCrosse 2009c).

So What: The Impact of Beliefs and Ethics on the Urban Environment

Three urban environmental debates coalesce around environmental ethics, science-based facts, and cultural beliefs: the impacts of residential lawns on water conservation and water quality, the cultural norms and environmental morals implicit in homeowner association landscape codes, and the ecological benefits and/or harm of using native and non-native plants in the landscape.

American neighborhoods are uniquely known for their perfectly (most of the time) maintained expansive green lawns. Cultural and environmental disagreements over the use of turf are perhaps the most contentious values versus science debates in urban areas. Empirical evidence points to overuse of water, pollution from chemicals, and loss of wildlife habitat as reasons to limit or eliminate the use of lawns. But historical, cultural, and economic values make losing the lawn anytime soon an unlikely proposition. Residential lawns have been a part of the American urban landscape ever since Frederick Law Olmstead, the father of landscape architecture, designed the first planned suburban community outside of Chicago in 1868. His vision for Riverside was to create the impression that everyone lived in a large park with lawns from house to house. In support of this vision, Frank Jay Scott (1870) wrote *The Art of Beautifying Suburban Home Grounds of Small Extent*, in which he noted that a smooth, shaven surface of grass was the most essential element of beauty, and it would be "unchristian" to hide this beauty from others. The popularity of lawns has been strong until now, when the lawn has become a moral battleground. For some, a lawn is a symbol of democracy, civic pride, and caring for nature; for others, a symbol of our anthropocentric environmental ethics and everything we do wrong with the environment (Pollan 1989). For some people, those who fail to conform to neighborhood norms of green turf are seen as lacking in civic values and responsibilities, and for others, those who do conform to the norm are seen as lacking in environmental values and concern for nature. This dilemma is especially true for those who live in homeowner associations, where home ownership usually means possessing or agreeing to values and beliefs that align with the codes and neighborhood norms. Ecology experts point to the overuse of irrigation water in urban areas as an argument to replace turf, or at least to adopt a "freedom" lawn, one that uses no irrigation or chemicals and allows for mowable weeds. Homeowners, however, often resist replacing their turf because they find it aesthetically pleasing, easy to maintain, economically important to their property value, and consistent with the neighborhood norms.

Several communities have adopted turf swap programs to decrease turf use by encouraging homeowners to trade turf for native plants. However, differences in perceptions and preferences have fueled an ongoing debate, primarily among ecologists and horticulturists, about the pros and cons of native plants and non-natives, including turfgrass. Non-native plants are often blamed for lack of biodiversity and becoming invasive in natural areas; however, evidence shows that non-native plants can be equally valuable food sources for some species, and some native plants are often seen as more aggressive growers in urban areas (Hoyle, Hitchmough, and Jorgensen 2017, 50). In addition, little evidence exists to support the notion that native plants improve soil, clean water, or mitigate the urban heat island any better than non-native plants. The preference for using non-native plants is related primarily to the aesthetic appeal of more colorful plants and perceived adaptability to climate change. Some communities have mandated a percentage of native plants be used in parks; however, this policy may be at odds with the preferences of citizens for the use of non-native plants, which could make long-term goals for sustainable urban policies more difficult to achieve (Hoyle, Hitchmough, and Jorgensen 2017, 62).

Now What: Ethical Perspectives for Urban Development

Several new ecological and environmental ethics philosophies have emerged to challenge some long-held environmental beliefs. A short review of the history of environmental philosophies is useful to understand how these philosophies materialized. In 1949 Aldo Leopold was the first to write about the importance of ecosystems as biological communities that sustain us, and about the moral duties of humans towards nature. In 1984 Arne Næss, a Norwegian philosopher, proposed the philosophy of "Deep Ecology" to express a vision of humans' relationship to nature, stating that humans are a part of earth and have no more rights than other forms of life. Key elements of this philosophy include rejecting the term "natural resources" because it assumes that elements and organisms are only economic commodities (Warner and DeCrosse 2009d). The emphasis is on the intrinsic values of nature, reduction in the human population, and a radical change in attitude toward consumption of resources. Ecologists who follow Deep Ecology argue that we should not continue to live in our conventional anthropocentric way, but instead explore other means of survival. The argument is for science, technology, and politics around the world to be transformed to respect the entire biotic community (Mokori and Cloete 2016).

Respecting the biotic community or ecosystem has contributed to the philosophical idea of holism: the idea that systems have certain properties that are evident only by considering the interrelationships of the components. This means that entire systems have moral significance and should be protected, or at the very least managed. Criticism of Deep Ecology and holism is the difficulty of applying and living out this philosophy in a practical manner. Finding the best way to integrate and apply this concept to a vision of how we should live in order to conserve ecological processes is the new frontier of environmental ethics. One suggestion for a new way of thinking is the concept of "relational" ethics—an ethic that is context driven and recognizes human/plant/animal interconnections and coexistence. This mutual dependence-based ethic means we are guided by the idea of care, where we act with respect and concern toward nature, acknowledging our gratitude and kinship, but also our differences, in our actions toward nature. For example, an ethics of care and respect means actively accommodating the needs of nature by creating urban forests and wildlife corridors, using restoration and conservation strategies in land use regulations, and changing transportation and energy systems to create a more sustainable city (Byrne 2011, 70–71).

A similar paradigm called environmental pragmatism has been proposed as a more workable philosophy that takes the pluralist perspective of valuing the environment in many different ways (Warner and DeCrosse 2009d). Several authors and philosophers have suggested that environmental ethics has failed to develop its practical task and is stuck on theoretical debates that prevent it from being practical. Environmental pragmatism is a new strategy that goes beyond theory to connect the methods of traditional American pragmatic thought to more practical and useful ways of solving real issues. Others argue that this new strategy or position is not a proper philosophical position since philosophy is an effort to clarify problems; others say it appears that philosophers are simply savoring a problem and not finding practical solutions because the usefulness of a theory should be based on its practical implications (Mokori and Cloete 2016, 5). Environmental pragmatists consistently endorse weak anthropocentrism as the value system most likely to motivate people to protect the environment, saying intrinsic value is too abstract to be useful and anthropocentric value (value to humans) appeals to an overwhelming majority of people. However, ordinary people can and do ascribe intrinsic value to nature, usually through a spiritual or religious connection (many believe it is wrong to abuse nature because God created it or they feel God's presence in nature). They may also have a nature ethic, or a belief that nature has rights. Intrinsic value can also be expressed through emotions such as love and awe, which are as powerful in motivating people to action as intellectual reasons. Because intrinsic value is a meaningful concept to make sense of our relationship with nature, it can be used by pragmatists to motivate people to a practical end (Callanan 2010, 133–138). The framework of pragmatism can also help decision makers and practitioners to integrate diverse values when considering a course of action. Multiple values will also help to reform practices and institutions by creating new policy discussions with new perspectives. Finally, pragmatism can be viewed as a system of many values that is flexible enough to unite traditional ethical values, diverse environmental ethics, and emerging values in a participative approach to decision-making (Mokori and Cloete 2016, 7).

Citizen Science Project: Poo Power! Global Challenge

Poo Power is a citizen science project to map dog waste hot spots to improve sustainability in public spaces and parks. Dog waste in cities contributes to pollution of waterways and creates unhealthy conditions in parks and neighborhoods. Cities around the world have tried awareness campaigns and the threat of fines to encourage dog owners to pick up after their pets, with varying degrees of success. The Poo Power iPhone App can be used to map dog waste "hot spots" to determine where they are located. The goal is to use the Global Poo Map to encourage discussion in classrooms about the social, scientific, and environmental effects of pollution and to make recommendations to local governments about where the hot spots are located and possible ways to solve the problem. The project also provides a good opportunity to discuss environmental ethics and the moral responsibilities of allowing pets to pollute public natural areas and cause possible health hazards (Cobb 2020).

Project: Largest Wildlife Crossing in the World

California is planning to build the largest wildlife crossing in the world to connect two large conservation areas. The Liberty Canyon Wildlife

Crossing will be a 200-foot-long bridge spanning the six-lane 101 Freeway in Los Angeles to help wildlife, such as mountain lions, coyotes, deer, lizards, and snakes move more freely between the sprawling urban landscapes. A major goal of the crossing is to connect two large populations of mountain lions that are isolated by the freeway in order to curb inbreeding and improve genetic diversity. The $87 million bridge will be funded by private sources (80%), and public funds allocated to conservation campaigns. The bridge will be the first crossing built near a major city and is proposed to be 165 feet wide with extensive plant cover and sound-and-light-blocking barriers so animals will not realize they are on a bridge. The design will cover about 8 acres total, with the bridge being about 1 acre. The proposal has nearly universal support; almost 9,000 comments on the draft environmental impact project had only 15 in opposition of the project (Solly 2019).

22

Valuing Ecosystem Services
in Urban Areas

Valuing ecosystem services (ES) of nature is an approach many cities are using in their planning and development process. In the urban setting, the primary purpose of defining ecosystem services and linking to an economic or social cost, or both, is to guide policy decision-making for urban growth, and to better integrate impacts on the natural environment in decision-making. For example, tree mitigation policies that require developers to protect and retain large heritage trees and replace all trees removed from the site provide for a future urban forest, protecting the ecosystem services provided by trees. When combined with other protection and mitigation policies, such as soil and water protections, a comprehensive ecosystem approach can be implemented. In the context of policy appraisal, valuing ES can help with determining whether a certain intervention that alters the ecosystem, such as land development, delivers a net benefit to society. Valuing ES can also provide evidence for decision-making on questions such as "value-for-your-money" and prioritizing funding. ES is also useful for choosing between possible best uses, for assessing liability for damage, and communicating to the public and to land managers the value of the environment (Defra 2007, 3, 13, 45–48).

Considerable research and several valuation processes have been applied to determine the value of the world's ecosystem services. Projects such as The Economics of Ecosystems and Biodiversity project (TEEB) and the Natural Capital Project (NatCap) have focused on large-scale wildlands and citywide scale, while small-scale urban green spaces have received less attention in the valuation process. One example is valuing residential landscapes that embody strong cultural and social values in addition to ecological services. Cultural services, such as aesthetics and health, are the most difficult services to value; nevertheless, residential landscapes are one urban green space that has a clear market value attached to it by the real estate industry. The value of a home landscape is typically considered in the context of return on investment (ROI) rather than on ecological benefits, especially for homeowners contemplating yard improvements. Studies show that landscaping affects market value for homes, and a good landscape can increase value by 5% to 15%, while poor or overly elaborate landscapes can decrease the value by 8% to 10%. Large trees can add anywhere from $5,000 to $10,000 to property value based on aesthetic appeal, but when regulating and supporting services are added, estimates are much higher. One reason that ecosystem services are rarely considered in residential landscape value is the lack of information on ecosystem services provided by residential landscapes to residents. Little data are available on landscape ecosystem functioning at the residential site scale, and suitable methods to apply the ecosystem service approach at the landscape scale have not been identified (Cook, Hall, and Larson 2012; Müller, de Groot, and Willemen 2010, 1, 7).

Natural resource experts involved in the process of valuing ecosystem services and creating policy include urban ecologists, conservationists, park and resource managers, environmental policy and law experts, and urban planners. Other areas of study related to resource use include economics, water and wetlands resource management, and fisheries, forestry, and wildlands management. Citizens also have a vested interest in valuing ecosystem services. The type and size of developments in their community, and the number and quality of parks and other natural resources are often dependent on development policies that protect and preserve natural areas and mandate the restoration of natural elements that have been destroyed. Many citizen science projects are linked to the concept of the value of nature. For example, the Natural Product Drug Discovery project looks for fungi in soil samples for drug discovery, an ecosystem service linked to human health and well-being. Biodiversity is a central tenet of ecosystem services, and several projects include mapping of biodiversity. In Los Angeles, California and Melbourne, Australia, projects include recording biodiversity indicators in different parts of the city. Additional projects include mapping and observation of land use change to detect possible losses of ecosystems and related ecosystem services.

What: Methods for Valuing Ecosystem Services

Valuing of urban nature is complicated by the mix of social, environmental, and economic benefits and competing interests among ecosystem users, plus the wide variety of stakeholders, including citizens, developers, administrators, and plants and wildlife (Chiesura and Martinez-Alier 2011, 93). Other urban considerations that complicate valuation include carrying capacity (the number of people and other living organisms a region can support without environmental degradation) and resilience of human-made ecosystems. Risk assessment and climate change, predictability and uncertainty due to constant change in cities, and equitable distribution of benefits are also part of the valuation mix. Urban policy based on ecosystem services requires a systematic approach that accounts for the impacts on ecosystems and links ecological effects to human welfare (fig. 22.1). The *Introductory Guide to Valuing Ecosystem Services,* written by the UK Department of Environment, Food, and Rural Affairs (Defra), outlines key steps for an impact pathway to policy change, including establishing an environmental baseline and identifying and providing a qualitative assessment of potential impacts of different policy options. Recommendations also include quantifying the impacts of options on specific ecosystem services, assessing the effects on human welfare, and valuing the changes in ecosystem services (Defra 2007, 4). Key issues include linking to scientific analysis that forms the basis for valuation and taking into account both the use and non-use values that individuals and society gain or lose

FIGURE 22.1. Portland, Oregon: a curb plaque reminds residents that green streets concepts, including no dumping of waste products that impact ecosystem and human health, are used to manage and clean stormwater.

when changes are made to natural systems. Use values include resources extracted from the ecosystem such as food or timber, nature-based recreation opportunities, and benefits supported by a resource such as climate regulation or pollution filtering. Non-use values include bequest value (value is passed on to future generations), altruistic value (making the resource available to others), and existence value (derived from the belief in the value of a resource), such as people donating to save elephants even if they will never see one (Defra 2007, 31).

Payments for Ecosystem Services (PES) is one means of acknowledging the economic worth of ecosystem services; however, the feasibility of using PES to support ecosystem management in urban areas has only recently been considered. PES is typically used to encourage individuals to avoid harmful environmental practices and contribute to the maintenance and preservation of services through a policy that provides a system of payments or incentives. Sometimes payments can come from beneficiaries of the service to its maintainers, such as paying to use a community park, but services are typically funded by governments and nongovernmental organizations (NGOs). Typically, methods in two different categories, economic and noneconomic, are used to determine value, depending on the application. Methods include:

- productivity: valuing services that help produce commodities, such as an aquaculture pond that produces fish for eating;
- hedonic pricing: considering services that affect the value of other commodities, such as a nearby forest that increases property value (fig. 22.2).
- travel cost method: measuring the value of a recreation area by determining how much people are willing to pay to travel and visit the site;
- damage cost avoidance, replacement costs, and substitution costs methods: for example, paying landowners to change management practices to prevent water pollution;
- the contingent valuation method: determining value based on asking people to choose between services, for example mowing the lawn or using mulch; and
- the benefit transfer method: estimating the value of services based on a completed valuation of a similar place (CEEweb n.d.; Holzman 2012, 6).

FIGURE 22.2. Gainesville, Florida: a residential lot contiguous to a forest provides a natural conservation buffer and increases the value of the property.

One or more of these methods can be used in development of urban policy that regulates the process of planning, permitting, building, and selling of properties. Currently various environmental planning and development policies consider the value of nature through tree mitigation policies, soil and water protection strategies, and recreation and cultural opportunities. More recent policies include climate regulation through tree planting programs, creation of conservation buffers, and wetland jurisdiction boundaries.

So What: Unintended Consequences of Putting a Value on Nature

Moral and ethical implications are often debated in the wisdom of putting a value on nature, including cultural and social values. Heritage sites, sacred sites, and territorial rights can be undermined by translating these cultural identities into a monetary value. Water privatization and water rights are examples of fairness and equitable distribution policies that create greater levels of inequality and disrupt communal social values. When base parts or elements of an ecosystem are commodified, it can impact the surrounding environment. For example, putting a high monetary value on trees for timber and farmable soil can lead to the destruction of forests and clearing for agriculture. The loss of the trees essentially means the loss of the entire ecosystem and provisions of multiple services that come from biodiversity.

Putting a value on nature is sometimes equated with the commodification of nature, meaning natural elements are incorporated into economic systems for a profit. However, there are several issues with this concept, one being the practical consideration of whether nature can be made into a commodity, another being the moral and ethical implications of commodification, and finally, the consequential effects of commodification on nature. Water is an example of a natural element that is not easily commodified because of its physical qualities; that is, the natural water cycle is a process without boundaries or clear property rights that make divisibility and exclusion difficult. In spite of this, water, which used to be completely free, has been turned into a privatized resource in many parts of the world where providing drinking water has become a multibillion-dollar economic industry (McWhinney 2019). Commodities need to be owned in order to have value and be traded, which means privatizing natural elements that were once communally owned, placing a natural resource in private hands and creating issues of access and equitable distribution.

Inherent challenges exist in valuing ecosystem services. The interdependency between and within ecosystems makes it difficult to disentangle certain services. A policy that impacts one system or part of a system may have unintended consequences for other services, and assessing for one service may not take into account other services. Spatial context is also a consideration as ecological boundaries and social boundaries rarely match; for example, water usually flows across many boundaries, so water quality may be more dependent on upstream areas outside the policy site boundaries. Temporal issues also need to be considered; the costs and benefits of services may be distributed over a long period, thus a "discount" rate for services, which converts all costs and benefits to present values, is normally used for comparison. An alternative is to estimate for the duration over which the benefits and cost will exist. Services also depend on the quality of the ecosystem; if the ecosystem deteriorates, the services will diminish and may disappear altogether. This concept is known as environmental limits (diminished services) and environmental threshold (loss of services). Cumulative impacts are also a concern: if decisions are made repeatedly and independent of each other, the total value of resources may be lost. For example, a decision might be made to allow development in an area with wetland habitat that makes up the bulk of the ecosystem service value per acre of land. If more development is allowed over time, each decrease in habitat size will reach a point at which the remaining wetland acres become worthless as habitat, with little ecosystem value. Other challenges include impacts that are not easy to value; for example, ecosystems may provide insurance value against climate change by increasing biodiversity to improve resilience and stability. In this case, underlying scientific evidence is needed for an overall analysis of value. Scientific evidence is also needed to help predict uncertain future losses of ecosystem services; however, even among experts there exists a lack of understanding about some aspects of services, particularly when nonlinear changes occur or when ecosystems respond in an unknown manner. The last and probably most important challenge is the scientific understanding of ecosystem function. More reliable predictions of the biological, physical, and chemical impacts on ecosystems are needed, especially in urban areas, including a better understanding of the cumulative effects and how systems compensate, and how to deal with lags in the development of noticeable impacts (Defra 2007, 41–42, 45–48).

Now What: New Approaches to Valuing Nature and Ecosystem Services

Payments for Ecosystem Services (PES) was discussed as one means of acknowledging the economic worth of ecosystem services with noted limitations of the system in urban areas. This has led to an innovative concept of *Payments for Urban Ecosystem Services* (PUES) that takes into account the different scales and configurations of green spaces in cites, the variety of public and private beneficiaries and providers, and the goals for ecosystem services in cities. The PUES system is a possible workaround for using PES in urban areas as summarized here (Richards and Thompson 2019). Landowners in urban areas include private commercial landowners (such as parks with sports facilities); private noncommercial landowners, such as homeowners; and public landowners, including city governments and agencies. Each group will have different motivations for changing their management practices to improve ecosystem services. Private commercial landowners want to maximize income potential, so PUES schemes would have to guarantee no loss in income and incentivize switching to management practices that could provide greater ecosystem services, such as planting landscapes that increase biodiversity and pollinator habitats. Homeowners are typically motivated by increased profit when selling their house, so they might participate in different management practices to increase ecosystem services that would reduce maintenance costs and increase aesthetics and recreation potential. They can also be motivated by activities that improve ES through incentives, such as tax rebates, for eliminating turf and irrigation systems. Public landowners, such as city parks departments, are motivated to reduce maintenance costs and provide public services that contribute to cultural ecosystem services. Public green spaces often have low economic return, with limited budgets for management practices that could improve delivery of ecosystem services, so it is important to identify stakeholders and the type of goals that would encourage investment in PUES. Schemes they could employ, such as charging admission to parks, should cover direct costs and help subsidize costs in areas where charging admission is not feasible to ensure equity in cultural services.

Practical and ethical considerations for designing and implementing PUES schemes are unique to urban areas. Fragmentation and high land values in urban areas will make it hard for PUES schemes to create large areas of intact ecosystems, so the urban form will require individual PUES that are small in scope and size. However, they can be scaled across large numbers of providers to improve ecological functioning, such as creating new micro-ecosystems like pocket parks and stormwater ponds. PUES should not increase social inequity; lack of equity can be affected by inability to participate in a scheme, wealth and power disparities, and inequity in the way payments are disbursed. For example, green spaces are often concentrated in wealthier neighborhoods and PUES may be more common in those areas, but new opportunities must be devised for low income areas. PUES should also reflect the many different values across different stakeholders; values associated with urban nature could be intrinsic, instrumental, or relational (values related to cultural identities or traditional practices). And finally, PUES must be grounded in science. Beneficiaries are more likely to invest in a scheme if they can predict the benefits they will gain and have the ability to verify the success of the scheme as it progresses. PUES will also need to quantify ecosystem services provided by different human-made urban ecosystems under different management practices, especially in under-studied areas such as the neighborhood, street, and residential lot (Richards and Thompson 2019).

Citizen Science: L.A. Nature Map

Biodiversity is a cornerstone of ecosystem services, but many communities don't have extensive knowledge of their urban nature. The L.A. Nature Map is a citizen science project that encourages residents to photograph and geolocate plants and animals in the Los Angeles area to contribute to the world's knowledge of L.A.'s biodiversity. The app lives in both the Nature Lab and the Natural History Museum of Los Angeles website. Observations are pulled from iNaturalist and displayed on an interactive map. Visitors can enter their address or zip code or the name of an animal to see which species have been observed near their location or throughout the city. The map shows over 200,000 observations of about 7,400 species (Ordeñana 2013).

Project: Evaluating Social Benefits with Public Life Tools

Social benefits of ecosystem services are the least studied and typically hardest to quantify. Environmental and economic benefits can be measured, while social benefits are observable but not easily measured. Social benefits can include user experience and satisfaction, improved health and decreased stress, educational experiences and overall improved quality of life (Yang, Li, and Binder 2016, 319). To help solve the problem, the Gehl Institute maintains a website, *Public Life Tools*, that provides a set of simple direct observation and survey tools to evaluate site use in public spaces such as parks and streets. Tools include age and gender tally, people moving count, quality criteria, participant surveys, and stationary active mapping. The tools are designed to use before and after the design intervention to evaluate changes in site use. The institute is also developing the Public Life Data Protocol, a set of metrics that are important to understanding public life with the goal of establishing a common format for collection and storage of data (Gehl Institute 2017).

23

Landscape Performance Data

Urban planning is a complicated process, with many jurisdictional layers and environmental concerns connected to development decisions. Stakeholder demands, codes and policies, environmental protection, and economic constraints are just a few of the issues that must be considered when designing and implementing large urban projects. As support from federal and state governments continues to decline, responsibilities for growth and development have been passed to local governments where decision makers, owners, investors, and policy makers are asking for more evidence that proposed projects will perform and provide a return on their investment (ROI) (LAF 2018, 1). City governments are increasingly requiring data to help determine ROI for projects as they are forced to be innovative in their search to raise capital. Funding strategies include private and philanthropic resources, leveraging public assets they own, levying special taxes, and using referenda and ballot initiatives to approve infrastructure investments. The key to making these strategies work is having a vision that is supported by data and analysis and a vision for growth that is unique to the community (Katz and Nowak 2018). Landscape performance data is a relatively new database used by planners and designers to measure ROI benefits and feasibility of green space projects. Landscape performance is defined by the Landscape Architecture Foundation (LAF) as "the measure of efficiency with which designed landscape solutions fulfil their intended purpose and contribute to sustainability" (LAF 2018, 1). Data about the environmental, social, and economic benefits of a project are used to determine landscape performance. The goal is to produce a body of knowledge that will help cities reduce investor risk and inform public policy. Typical projects for evaluation include planned communities, urban parks, green streets, campuses, schoolyards, greenways, and restoration projects (LAF 2018, 3).

Performance is based on research by a wide range of disciplines, including landscape architecture, ecology, horticulture, civil engineering, transportation planning, public health, and economics. Urban landscape ecology principles, in particular, provide a foundation for performance-based design. Professionals who work with performance data include landscape architects, planners, city maintenance crews, tourism offices, developers, city officials, and urban ecologists (LAF 2018, 1). The development process offers many opportunities for citizen science activities such as conducting a BioBlitz, a one-day activity documenting plants and animals on the site before and after construction, described in chapter 18. Citizen science projects with crowd-sourced databases such as iNaturalist, i-Tree Eco, and eBird, are often used as tools for data collection. For example, a project with a goal of increasing bird species by improving habitat might use eBird to document bird species, iNaturalist to document and geolocate plants and animals, and i-Tree Eco to geolocate and document all trees on site. The information can then be used to develop a site plan that includes adding wetland areas for wading birds, and a planting plan that includes increasing native tree species and planting more shrub species that attract insects for food.

What: Performance Measures for People, Economics, and the Environment

The Landscape Architecture Foundation (LAF) maintains the Landscape Performance Series website with case study briefs and resources, such as training guides and toolkits, to help practitioners collect landscape data on their projects. The key to using performance measures requires selecting the appropriate metrics (data to be collected) and the appropriate method (the means to quantify the metric), for both collection and evaluation. Each category—social, environmental, and economic—will have different but related metrics with the common goal of sustainability, which requires a set of measurable sustainability indicators. For example, social improvement metrics might include the number of people using alternative transportation, number of citizens reporting higher quality of life satisfaction scores, and public health data showing reduced risk of disease and poor health caused by poor water or air quality. Evaluating performance requires measuring outcomes, such as benefits, rather than outputs, such as number of trees planted or miles of trails. For example, the number of

FIGURE 23.1. Placerville, California: repeated trimming of tree to expose parking lot light is a sign of a poorly performing landscape.

trees planted is based on stated goals, including desired amount of carbon sequestered or desired temperature reduction, which are the measured benefits or outcomes (LAF 2018, 3).

Methods used to collect data include existing public data sources such as tax or utility records, data from citizen science projects, and rating systems such as Leadership in Energy and Environmental Design (LEED) or Sustainable SITES Initiative (SITES). Research methods such as surveys, field studies, and field observation are also used to collect site data. For example, post-occupancy evaluation (POE) is one method of collecting data for social metrics. As the name implies, researchers use different techniques to collect evidence after occupancy on how spaces are used and how they meet the needs of the user. Use patterns and user preferences are often collected for urban green spaces designed for community activities. Site observation might include counting number of visitors and mapping movements, recording signs of wear such as trampling of vegetation and misuse of facilities, or noting repeated maintenance (fig. 23.1) and repair activities. Information from a POE is often used in the design process to create evidence-based designs for new projects (Yang, Li, and Binder 2016, 314).

Other methods are used to determine costs of building and maintaining projects to determine funding needs prior to construction or to show alternatives for cost savings that can be considered. Value landscape engineering (VLE) is a method used to consider the value of all the services associated with a landscape over its lifetime. The goal is to reduce inputs (such as water use and maintenance labor) while maximizing value (services). For example, landscape features such as trees are assessed on the cost to purchase and install, the cost of annual maintenance, and the cost to remove or replace. The services provided by trees such as CO_2 emissions and carbon sequestration are also measured during installation, maintenance, and removal (Rosenberg et al. 2011, 636–637). Life cycle assessment (LCA) is a method for determining the costs over the life of a product or materials used on the site. Environmental impacts associated with all stages are also considered, including extraction of raw material and the process of manufacturing, distribution, maintenance, and disposal or recycling of the product. Information is used to evaluate impacts associated with the inputs and releases (such as volatile organic compounds) through each of the life stages. LCA studies help to promote responsible design and reduce environmental impacts (SAIC 2006).

Methods to quantify the data include comparisons, reporting absolute and relative values, monetizing (assigning a dollar value), and projecting changes over time. Comparisons are the most widely used and effective way to measure impact, including before and after data on a given metric, comparing conventional design practices to sustainable practices, and using a benchmark or average for comparison (LAF 2018, 4).

LAF has developed a guidebook, *Evaluating Landscape Performance, A Guidebook for Metrics and Methods Selection, 2018* that lists appropriate metrics (data) for each of the three categories: environmental, social, and economic. Within each of these categories are subcategories that further define the metrics and suggest possible methods to quantify the data. The six subcategories for environmental benefits include land; water; habitat; carbon; energy and air quality; and materials and waste. Within these categories specific issues are listed for assessment; examples include:

- preservation of land and soil and preservation and restoration of habitat;
- improving stormwater management, water conservation, flood protection, and groundwater recharge, and
- improving air quality and carbon sequestration (LAF 2018, 15).

FIGURE 23.2. Gainesville, Florida: a restored pond in a historic district is an example of social, environmental, and economic benefits from nature that add to the value of the neighborhood.

Social benefits have 10 subcategories. Some are concerned with the value we place on land and others are concerned with livability. Examples of activities to be assessed within these subcategories include:

> restoring culturally significant features (fig. 23.2), supporting urban agriculture, and improving the visual quality of the area;
> fostering knowledge and awareness, promoting play and relaxation, and supporting physical and mental health by encouraging walking and biking; and
> reducing levels of undesired noise and reducing crime and perceptions of danger (LAF 2018, 51).

Economic benefits include seven subcategories: property value, operations and maintenance savings, construction cost savings, job creation, visitor spending, tax revenue, and economic development. Assessment items in this category include:

> adding value to the site and adjacent properties;
> reducing ongoing and one-time costs;
> providing employment during construction or ongoing operations;

generating revenue from visitors or through property and sales tax, and catalyzing real estate and business investment (LAF 2018, 75).

The categories were developed as guides to selecting metrics and methods that will fit individual project goals. The categories, metrics, and methods can be selected, modified, or adapted for individual projects, and they can be used to generate new metrics for unusual or unique projects. Since all categories and metrics will not be applicable to all projects, the list should be narrowed based on the design intent and goals. Categories can also be used as a guide to create project goals (LAF 2018, 7).

So What: Performance Data for Sustainable Design

Data and performance metrics are needed to show numerical values for the performance benefits of certain types of landscapes and make the case for more sustainable development. Planners and landscape architects face many challenges, including climate change and rapid urbanization, that require new ways of thinking about resilient landscapes (fig 23.3) and new processes for design and implementation. To meet these challenges, current design practices need to be evaluated and empirical evidence published to guide decision-making, inform public policy, and improve return on investment (Yang, Li, and Binder 2016, 315).

FIGURE 23.3. Mexico City Beach, Florida: a lost landscape is an opportunity to design a more resilient landscape to repair damage from Hurricane Michael.

Over time, as more case studies become available, the benefits of certain landscape practices will be proven through performance data that shows improved environmental and social health. While the data is useful for making informed decisions on future projects it can also be used for other purposes, including design inspiration, education, tourism marketing, social marketing, and developing new assessment tools and construction techniques. For example, social marketing messages based on landscape performance data can be used to change landscape maintenance behavior. Data on irrigation and water use in a community development compared with data from a similar development can inspire homeowners to regulate their own water use.

Current project certification programs such as Leadership in Energy and Environmental Design (LEED) and the Sustainable Sites Initiative (SITES) have assessment programs that evaluate preconstruction design strategies and have only recently added post-occupancy evaluation strategies. The challenge has been finding practitioners and researchers with the time, knowledge, and skills to assess a project on the quantitative level. The key is to make the process simple, but effective, by selecting methods and metrics that are easy to use for nonexperts, are applicable to a wide range of projects, can be measured in a short time frame, and are defensible (LAF 2018, 7). A variety of new online calculation tools are available to help with the process. A new research program is supported by LAF to quantitatively demonstrate the environmental, economic, and social benefits of a project through the review of many high-quality case studies.

Now What: Tools and Methods for Measuring Landscape Performance

Landscape performance data are useful to practitioners and several agencies and organizations. Many design firms have in-house evaluation programs which they use to improve their projects and design practices. City planning departments use the data from case studies to evaluate and revise development and construction policies for sustainability. Case studies are good examples for cities that are making decisions on similar projects, and they provide information to guide research on the methodology and categories used for evaluation. LAF has created a benefits toolkit—a searchable collection of online tools and calculators to estimate landscape performance. The best methods are actual direct measurements, such as water and air quality tests, air temperature records, number of users, mapping of use patterns, and tax and employment records showing actual economic gains. However, when direct measures are not available, tools to estimate performance can be used. Various agencies have developed online calculators to estimate such things as carbon footprint, annual costs for landscapes, services, and goods provided by nature, and benefits from green infrastructure.

The Resource Conserving Landscape Cost Calculator developed by the U.S. Environmental Protection Agency is used to compare the cost of converting conventional landscapes to landscapes that require less irrigation and grow more slowly. The calculator shows initial and annual costs for the current landscape and the planned retrofit (USEPA GreenScape Tools 2016). The Landscape Carbon Calculator is used to estimate carbon footprint and time to carbon neutral based on site design and management, with design suggestions to reduce the carbon footprint (Climate Positive Design n.d.). Another tool, the Universal Floristic Quality Assessment (FAQ) Calculator, is an open-source project to measure a site's habitat condition or the condition of a specified natural plant community's condition. Users upload an inventory of individual plants on the site using common or scientific names, and an Excel file of results is downloaded that summarizes conservation-based metrics (Freyman, Masters, and Packard 2016). The Natural Capital Project (Stanford University) developed InVEST, Integrated Valuation of Ecosystem Services and Tradeoffs, a suite of 18 software models for land, freshwater, marine, and coastal ecosystems used to value, and map, services and goods provided by ecosystems. The models calculate the potential of ecosystems to provide human benefits such as carbon sequestration, water purification, hydropower production, and offshore wind energy. A new model for urban areas is currently in development. The models can be used to compare scenarios in order to balance environmental and economic needs (InVEST 2016). American Rivers and Center for Neighborhood Technology (CNT) developed a report, *The Value of Green Infrastructure: A Guide to Recognizing Its Economic, Environmental and Social Benefits,* to help quantify and place an economic value on benefits provided by green infrastructure. The report presents simple equations to quantify water and air quality as well as climate change benefits from planting trees and using green roofs, permeable pavement, bioretention, and water harvesting (Center for Neighborhood Technology 2018).

Some work remains for the use of performance data in landscape design. Studies show that social benefits are increasingly important in landscape performance assessment, pointing to the need to integrate so-cial science theories and advanced research methods into the social ben-efits assessment. Ecosystems services is another source of information for a better understanding of the benefits of sustainable design. A theoretical and analytical framework must be developed to support long-term analy-sis in landscape performance assessment. Long-term analysis will help to identify patterns and trends to advance understanding of resiliency in landscapes and to identify contextual factors such as surrounding urban conditions and land cover in the performance data analysis (Yang, Li, and Binder 2016, 326).

Citizen Science Program: eBird

eBird is a database created by citizen science bird observations in the field using the eBird mobile app. Bird observers around the world are upload-ing the date, location, and number of birds sighted at a particular location to track bird species distributions and abundance worldwide. The online database is freely accessible, and location is the only required input to ac-cess the records of bird sightings. A rigorous eBird verification process is used to ensure data quality. The data are useful for understanding which bird species should be, or are likely to be, present in a particular region when designing to improve bird and wildlife habitat. The information can be used to determine which tree and shrub species should be used on the site (eBird. n.d.).

Case Study: Performance Data for Renaissance Park, Chattanooga, Tennessee

Renaissance Park is a 22-acre urban brownfield redevelopment project that transformed a contaminated post-industrial site into a public park. In 2006 the existing industrial structures were removed along with 34,000 cubic yards of contaminated soil that was sealed in several unusual land-forms on the site. Excavation of the contaminated soil increased the flood-plain storage by 15,000 cubic yards, and the void created an area for a con-structed wetland. The improved habitat value of a stream on site went from "marginal" to "suboptimal," and the USEPA Rapid Bioassessment scores increased from 60 in 2002 to 122 in 2014. The project was also able to re-duce irrigation demand by 74%, or 1.6 million gallons per year, compared with a baseline case with 79% turf. Social benefits were measured by user surveys of the estimated 145,220 annual visitors. Of the visitors surveyed, 81% agreed that the park increased their outdoor activity, and 89% said they shopped or dined within a half mile of the park before or after using the park. In a survey of visitors who live within one mile of the park, 41% said they were willing to pay a premium to live close to the park. Economic stimulus has improved in the area; since 2005, $55 million has been invested in redevelopment projects adjacent to the park, and five properties within a quarter mile of the park have been redeveloped. Property values have also increased in the area; the aggregate land value within a quarter mile of the park increased by 821% between 2005 and 2013, compared with increases of 319% in other properties in the neighborhood. The project builders were able to save a little over $1 million in construction costs by salvaging the concrete factory floor from the site and reusing it as fill, and per-acre labor maintenance costs were reduced by $4,500, or 73%, per year compared with an adjacent park of similar size with large expanses of lawn (Landscape Performance Series 2018).

24

Environmental Justice

Environmental issues provide us several opportunities to evaluate social equity, health, and safety in neighborhoods, districts, and entire cities. Environmental conscience developed more keenly in the 1960s and 1970s, when both citizens and governments finally accepted the fact that damage to the environment was detrimental to the health and well-being of humans. One of the first calls to action to seriously address environmental issues was Rachel Carson's book *Silent Spring*, published in 1962. In it, she related the environmental damage caused by pesticides and herbicides widely used in pest control and agricultural programs that were destroying natural habitats and wildlife, and causing yet unknown dangers to human life.

The harbinger of Carson's concerns was a colorless, tasteless, and almost odorless insecticide, DDT (dichloro-diphenyl-trichloroethane). This synthetic crystalline chemical compound was finally banned worldwide in 2001, but we are still living with its damaging consequences: children exposed to it are likely to develop cancer later in life; it causes permanent damage to the nervous system; and its breakdown products are still being detected in people (NPIC 1999). As with other environmental disasters, the dreadful effects of DDT on the environment were an unforeseen consequence of the post–World War II Green Revolution that put research technology to use to exponentially increase agricultural production and feed the ever-growing world population. Food production and distribution continues to be a challenge, and most nations in the world will continue to deal with the delicate balance between conserving their natural environment, urbanizing, maintaining enough farmland to feed their populations, and producing enough food without the use of toxic chemicals.

Silent Spring is credited with launching the environmental movement in the United States. Calling attention to the long-term damages caused by pesticides not only outraged people potentially affected by them, but also forced governments to acknowledge the need for changes in policy, including regulating the chemical industry that was greatly benefiting from the sales of harmful compounds. One of the questions raised by Rachel Carson was whether the public has a right to know about environmental issues that can cause human health risks—the same question still being asked today by environmental justice activists. More than 50 years after publication of her seminal book, it has become clear that the public has not only the right to know, but also the right to act. Communities are very much engaged in most environmental justice issues today, and grassroots movements are participating in community initiatives to ensure equity, health, and safety for all.

What: What Does Environmental Justice Mean?

Environmental justice can be considered from both an amenity and a hazard perspective. A pragmatic view of environmental justice considers the distribution of amenities and hazards across the territory. Several studies have analyzed environmental justice issues in communities from these two different perspectives, and the result is uniformly the same: where there is inequity, poorer and ethno-racial minority neighborhoods have less access to environmental amenities and are exposed to more environmental hazards, while affluent neighborhoods have more access to environmental amenities and are not exposed to environmental hazards (De Sousa 2006; Wolch, Byrne, and Newell 2014; Rigolon 2016). In addition to the distributional inequity of environmental ills and benefits, it is often the case that low-income, ethno-racial minority groups are not included in the decision-making processes that lead to the location of facilities in a given territory, whether amenities or hazards (Schlosberg 2004).

Green open space and urban parks are examples of amenities when considering environmental justice. Accessibility to natural areas and open green space has long been recognized as a public health issue; the impact of exposure to nature on healthy behavior and general well-being has been studied extensively (for example, Kaplan and Kaplan 1989; Ulrich et al. 1991; Bedimo-Rung, Mowen, and Cohen 2005; Ward Thompson 2011; Wolch, Byrne, and Newell 2014; Rigolon 2016; Eldridge, Burrowes, and Spauster 2019). Power plants, landfills, toxic-waste incinerators, water and sewage treatment plants, and various types of brownfield sites are examples of hazards when considering environmental justice (fig. 24.1). In the United States, brownfields are defined as "real property, the expansion,

FIGURE 24.1. Curitiba, Brazil: lack of basic sanitation and environmental hazards in close proximity to where the poor live.

redevelopment, or reuse of which may be complicated by the presence or potential presence of a hazardous substance, pollutant, or contaminant" (Public Law 107–118, H.R. 2869, p. 6).

Studies have shown that disadvantaged populations are disproportionately exposed to hazardous facilities (for example, Hollander 2009; Wolch, Byrne, and Newell 2014; Rigolon 2016; EPA 2019a). In the case of brownfields, their conversion creates a new opportunity to turn hazards into amenities and green the city in the process, creating more sustainable urban environments (Harnick 2000; De Sousa 2004, 2006). Because brownfields are often located in close proximity to low-income communities, most brownfield remediation projects attempt to give the area an economic boost; with this purpose, they are usually converted into commercial or residential areas. However, parks, greenways, and other recreational uses, albeit not producing an immediate economic return, can bring significant benefits to a city, including an equitable and just distribution of amenities.

So What: Environmental Justice in Action

The environmental justice literature evolved in the 1980s, following the initial movement toward environmentalism of the 1960s and 1970s. As with the movement, research at first focused on documenting the disproportionate exposure of low-income and ethno-racial minority populations to environmental hazards, such as brownfields. The focus on environmental amenities is more recent, and one of the most focused-on amenities is open green space. Most studies have analyzed the spatial distribution of urban parks, with particular attention to equity (or lack thereof) across income and ethnic categories (Wolch, Byrne, and Newell 2014; Macedo and Haddad 2016). Several cities are investing in their parks systems, not only as a way to improve their sustainability and environmental health through green infrastructure development, but also as a way to improve social equity through opportunities for recreation, physical activity, and healthy lifestyles (Eldridge, Burrowes, and Spauster 2019). Among the various environmental justice topics, spatial equity is also considered an important measure of sustainability (Pickett et al. 2011). Most sustainable cities try to offer equitable access to environmental amenities to most, if not all, of their population. Several studies assume equitable cities are those where green open space equally benefits every citizen (Wolch, Byrne, and Newell 2014; Macedo and Haddad 2016; Rigolon 2016; Rigolon, Browning, and Jennings 2018).

A positive outcome of broader participation and activism driven by the environmental justice movement has been the conversion of brownfield sites into green space assets as a planning strategy to increase green open space in several cities across North America (De Sousa 2004, 2006). The first U.S. law to regulate brownfield remediation was the 1980 Comprehensive Environmental Response, Compensation, and Liability Act (CERCLA), popularly known as Superfund. This law was amended in 2002 by the Small Business Liability Relief and Brownfields Revitalization Act, popularly known as the Brownfields Act (EPA 2019a). Both acts provide funds to clean up brownfields and eliminate substances that are hazardous to the environment and that may endanger public health. These acts have allowed the federal government to allocate funds to communities to assess and remediate former industrial sites and to convert brownfields into sustainable uses; nonetheless, states have borne the brunt of cleanups. The federal program has funded the assessment of more than 30,000 sites but has paid for the cleanup of only 2,000, while brownfield state funds have paid for the cleanup of 150,000 sites (EPA 2019a). In addition to restoring the environmental health of the site and its surrounding area, once cleaned up, brownfields can be transformed into open green space (Eldridge, Burrowes, and Spauster 2019).

One exemplary case of brownfield remediation is that of the Fairmount

Corridor in Boston, Massachusetts (Levine 2013). In this project, federal agencies, such as the Environmental Protection Agency (EPA), the U.S. Department of Housing and Urban Development (HUD), and the U.S. Department of Transportation (DOT), provided funding and technical assistance while the project was spearheaded by a collaborative that included three Community and Economic Development corporations, Dorchester Bay, Codman Square, and Southwest Boston. More than 500 brownfield sites were identified in these low-income, ethno-racial minority neighborhoods (https://www.mass.gov/brownfields-cleanup). In addition to remediation, the communities in this area did not have access to public transit, so another grassroots organization, the Greater Four Corners Action Coalition, got involved, and the Fairmount Indigo Initiative was launched (https://fairmountcollaborative.org). The government of Massachusetts, namely the Department of Environmental Protection, the Department of Transportation, and the Massachusetts Bay Transportation Authority, joined in the transit equity campaign, and the Boston Redevelopment Authority also partnered with the collaborative (Boston 2019). With community groups and agencies from federal, state, and local governments working together, it was possible to successfully achieve environmental improvement and economic redevelopment. Brownfield redevelopment is more successful in turning environmental remediation into economic development when interagency partnerships and community stakeholders work together to achieve desired outcomes (Eldridge, Burrowes, and Spauster 2019). From a policy perspective, cohesiveness of federal, state, and local policies is necessary to ensure success.

Now What: Making Urban Environments Just

The first step toward justice is recognition of injustice. If the public and the government had not recognized Rachel Carson's claims in *Silent Spring*, perhaps the decision-making process that led to the creation of the Environmental Protection Agency would not have occurred. Environmental justice can be defined in terms of both decision-making processes and their spatial outcomes; an equitable geographic distribution of environmental amenities and hazards can result only from fair decision-making processes regarding their location (Schlosberg 2004). Local communities need to have a say in this process, and they usually need assistance articulating their concerns. In addition to distributional aspects, other key components of environmental justice include participation, activism, and

FIGURE 24.2. Curitiba, Brazil: a former quarry converted into a park.

scholarship (Boone and Fragkias 2013). From the standpoint of environmental justice activists, it is necessary to "call for policy-making procedures that encourage active community participation, institutionalize public participation, recognize community knowledge, and utilize cross-cultural formats and exchanges to enable the participation of as much diversity as exists in a community" (Schlosberg 2004, 522).

One of the ways that environmental justice has been successfully achieved in terms of equitable access to open green space, for example, is to give all segments of the population—regardless of race, ethnicity, or income level—access to parks, forests, and natural amenities (fig. 24.2). This access needs to be achieved from different perspectives; it is not only a matter of quantity, but also of quality. In the case of urban parks, research has shown that both the acreage and the quality of parks in low-income, ethno-racial minority neighborhoods is less than those in affluent neighborhoods (Wolch, Byrne, and Newell 2014; Macedo and Haddad 2016; Rigolon 2016; Rigolon, Browning, and Jennings 2018). For environmental justice to be served, there needs to be parity in quality as well. In fact, one could argue that, to achieve true environmental justice, low-income neighborhoods should have better quality and more quantity of natural and recreational amenities with free access because their residents lack the means to frequent private clubs and gyms, while affluent populations have other paid options (Macedo and Haddad 2016). Social equity can be promoted in cities with increased access to environmental amenities such as urban parks and forests. This equity can be measured in terms of accessibility, quantity of green space (acreage), quality of urban parks (equipment, natural and built features), and safety.

Another approach to creating just urban environments is to focus on sustainable development (Haughton 1999). The ethics of environmental justice protect a broader community; it is not only the rights of disadvantaged communities that need to be preserved, but also the rights of all living creatures and those of future generations. Grounding sustainable development initiatives on ethical principles ensures that communities already dealing with preexisting vulnerabilities, such as long-standing social and economic inequality, can benefit from development in their communities. Improving the ecological quality of our cities in general may be a way to refocus the environmental justice discussion; rather than arguing over ways to distribute environmental hazards equitably, perhaps we should all be working to reduce, mitigate, and eventually eliminate them everywhere.

The environmental justice movement has successfully united disparate issues and experiences of injustice (Schlosberg 2004). Environmental justice movements cannot be uniform, because they are grounded on distributive justice and, as such, they challenge the discourse of development. Gentrification, for example, can be a threat to social justice; unintended consequences of actions taken to create a more equitable distribution of amenities and hazards in the urban space may result in exclusion of the disadvantaged. The implementation of parks and other natural amenities, the remediation of brownfields and other environmental cleanups, and multiple initiatives to increase the sustainability of cities and to improve their ecological quality may have negative outcomes from a social justice perspective. Making urban environments just requires an ethical approach to environmental issues and wise ecological stewardship.

Citizen Science Program: EPA Science Tools

The U.S. Environmental Protection Agency (EPA), through its Office of Research and Development (ORD) Sustainable and Healthy Communities (SHC) program, supports and manages numerous citizen science projects. These projects provide unique opportunities for the EPA, sometimes working in collaboration with other federal agencies and nongovernmental organizations, to engage with the public in environmental protection initiatives to better understand and solve environmental problems and to advance the agency's mission. One example is the partnership between EPA's Region 10 and several communities across the Pacific Northwest to introduce academic institutions, community-based organizations, and state and local governments to the EPA's Community Focused Exposure and Risk Screening Tool (C-FERST), a web-based environmental information and mapping tool. C-FERST allows community members to take on the role of scientists and identify and assess potential risk and exposure to environmental hazards at the local level. The data collected by both the community and agency, including maps, photographs and notes, are incorporated into an EnviroAtlas Interactive Map. The EPA Region 10 project is part of the Regional Sustainable Environmental Science (RESES) research program. It uses colleges and universities to engage with disadvantaged communities and address their environmental concerns, which range from brownfields to community livability to urban environmental education. Specific examples include children's health and asthma along the I-5 corridor in Portland, Oregon, and healthy food systems in Tacoma, Washington (GSA n.d.).

Case Study: Urban Parks in Curitiba: From Sanitation Infrastructure to Ecological and Social Assets

Unplanned urban development and the rapid increase of impervious surfaces, added to the channelization of rivers to facilitate the construction of urban infrastructure, made floods one of the most serious urban problems in the city of Curitiba, Brazil. The worst flood on record occurred in 1932 when the Ivo River rose 6 feet above level, flooding most of downtown. The first public park was created as part of the city's first large sanitation project, the channeling of the Belém River (fig. 24.3). This relatively small park, about 70,000 square meters (17 acres), was created in 1886 in a swampy area near the fledgling business district. This project set an important precedent, and virtually every park created in Curitiba since then has been designed to mitigate floods. These parks not only allow for vegetated areas to be interspersed in the urban fabric, but every park also has a retention pond for flood mitigation and water filtration purposes. Riparian areas in need of restoration or conservation, flood-prone and low-lying areas, and swampy terrain have always been preferred for the implementation of parks in Curitiba.

FIGURE 24.3. Curitiba, Brazil: Passeio Público, the first urban park created in the city and the only one downtown.

Today, Curitiba has 27 parks and 16 wooded areas, totaling about 23 km² (5,700 acres) in an urban area of 432 km² (167 square miles). Most of these areas are located in the northern part of the city, where elevations are higher and slopes steeper; these are also the areas with most natural beauty. Some parks were created in "recycled" spaces, for example, brownfields and abandoned quarries that offered no other feasible development opportunities. Most park development in Curitiba has been enabled by legislation enacted in the 1990s, which not only permitted, but also encouraged transfer of development rights for conservation of green open space. Curitiba was one of the first cities in Brazil to implement policies to allow transfer of development rights for historic and environmental preservation. The precepts of this isolated legislation were integrated into the 2000 revision of the city's master plan, specifically directing the funds from transfers of development rights to environmental conservation and social housing. The impact of this legislation is measurable: 17 of the 43 parks that exist in Curitiba today were created during the 1990s, and an additional 11 since 2000 (Macedo 2013; Macedo and Haddad 2016). The implementation of these parks has prevented floods in addition to creating recreational opportunities for the population.

25

Healthy and Sustainable Neighborhoods

A healthy and sustainable city needs healthy and sustainable neighborhoods to offer a better quality of life for its citizens. Neighborhoods are the building blocks of a city, and by making our neighborhoods sustainable, we can achieve citywide sustainability. Most sustainability indicators are based on social, environmental, economic, and institutional aspects of sustainable development. Successful redevelopment projects to make areas more sustainable usually take place at the neighborhood level, since it would be impossible to make wholesale changes to entire cities. Each neighborhood, each community, plays a vital role in creating healthier and more sustainable cities and metropolitan regions.

Community buy-in and citizen participation are requirements for sustainable actions. Bottom-up approaches to city planning generally start at the neighborhood level. Invariably it is more feasible to engage with communities at the neighborhood level rather than at the city level; however, there are initiatives that directly affect neighborhoods but necessitate municipal-level action. For example, when a transit system is developed or extended in a city, it may have a major impact on neighborhoods, and yet, that scale of infrastructure demands integration of city policies and neighborhood needs. Some traditional neighborhoods have been divided and destroyed by transportation infrastructure when only municipal or regional interests were taken into account (Dluhy, Revell, and Wong 2002; Connolly 2006). By the same token, there are services and activities that are better implemented at the neighborhood level. For example, some communities are experimenting with block- and district-size waste management and energy generation and sharing through blockchain technology (Green and Newman 2017; Green, Martin, and Cojocar 2018).

FIGURE 25.1. Vancouver, Canada: community gardens provide fresh organic produce, contributing to the health of urban populations.

The city needs to determine clear targets for its communities; however, it is at the neighborhood scale that the most creative design solutions can be implemented (fig. 25.1). Neighborhoods can be made healthier through the implementation of green infrastructure. In addition, neighborhoods can more easily catalyze initiatives and actions that may impact the entire city; community gardens are an example.

What: What Makes Communities Healthy and Sustainable?

Communities can be made more sustainable and allow their residents to lead healthier lives by making changes to their built environments. The kind of urban environments created in the United States, particularly during the second half of the twentieth century, have not been very conducive to healthy living. Americans traveling in Europe often wonder why we cannot have a similar lifestyle in the United States (Rybczynski 1995). We tend to enjoy compact urban forms and the diversity of European urban environments; however, when we return to our routines, we are not willing to replicate the healthy behaviors we adopt while on vacation.

Two different eras of urbanization directly impacted the health of urbanites in the United States (Hall 2002). The first was at the end of the
nineteenth century, when the industrial revolution caused urban centers
to rapidly grow and densify, concentrating labor in cities. The densification
of cities was problematic because they lacked the infrastructure necessary
to absorb the sudden and intense population growth brought by indus-
trialization. Corrective measures were introduced by the sanitary reform
movement in the late nineteenth century and the importance of green
spaces in cities was recognized (Olmsted 1870) (fig. 25.2). The second era of
urbanization was the result of post-Depression policies in the 1930s combined with postwar programs and incentives (Radford 1996; Hall 2002;
Rose 2009). The one program with the greatest and most long-lasting impact on American built environments was the Interstate Highway System
(Hall 2002). Introduced by President Eisenhower after World War II, the
U.S. Interstate Highway System allowed cities to be permanently evacuated
daily rather than only in the event of a war, as General Eisenhower had
idealized. Our "freeways" gave everyone the freedom to live far away from
the city core; essentially, the Interstate Highway System subsidized the
suburbanization of America. The subdivision of inexpensive land brought

FIGURE 25.2. New York City: Central Park, conceived and designed by Frederick Law
Olmsted in the late 1800s, provides refuge, respite, and infinite opportunities for physical activity.

development and irreversible sprawl to every metro region in the country. A new urban typology, dubbed "edge cities" (Garreau 1991), became ubiquitous. Edge cities are urban areas that comprise most urban uses—residential, commercial (big box retail and office parks), and institutional (schools and libraries)—however, they are not cities. They usually sit in the unincorporated areas of counties, near the intersection of two or more interstate highways, nameless places without city limits, local government, or any other feature that would characterize an urban community.

Urban sprawl has had significant impact on American landscapes and has presented a real threat to sustainable development efforts. Generally, sprawl is characterized by low-density development; neighborhood layouts that lack connectivity and create automobile dependency; no core and few public green spaces; Euclidian separation of uses with a limited number of housing types; and no acknowledgment of the public realm (Ewing, Pendall, and Chen 2002; Gillham 2002). Suburbs, edge cities, and exurbs are different types of sprawl that have the same detrimental effect on the landscape and perpetuate unhealthy and unsustainable behaviors.

So What: Unforeseen Consequences

The suburbanization of America was the result of well-intended policies that, at the time they were enacted, brought economic development and prosperity to most urban areas. Fifty years later, we recognize the unintended consequences of this particular form of urbanization, and now we need to develop strategies to solve the problems it caused. Unfortunately, American sprawl was disseminated as a desirable style of development, and the same problems are being perpetuated in other countries in the world (Schmidt 2004). Namely, Australia's four largest cities, Brisbane, Perth, Melbourne, and Sydney, are the top four most sprawling cities among world metro regions, followed by Calgary and Vancouver in Canada.

The exodus of population, especially the more affluent, to suburbs and exurbs has caused the decline of urban cores; central areas were depreciated and, in some cases, turned into blighted, poverty- and crime-ridden places (fig. 25.3). Downs (1999, 956) identified 10 criticisms of sprawl: "(1) unlimited outward extension of development, (2) low-density residential and commercial settlements, (3) leapfrog development, (4) fragmentation of powers over land use among many small localities, (5) dominance of transportation by private automotive vehicles, (6) lack of centralized planning or control of land uses, (7) widespread strip commercial development,

FIGURE 25.3. Paraná, Brazil: exurbs and other forms of American sprawl appearing in many other cities in the world.

(8) great fiscal disparities among localities, (9) segregation of types of land use in different zones, and (10) reliance mainly on the trickle-down or filtering process to provide housing to low-income households." These criticisms and their detrimental effects have been confirmed in cities all over the United States. Some have successfully revitalized their downtowns, while others are still lingering (Ford 2003). In addition to social problems, sprawl has caused serious environmental damage, with repercussions for health.

Most suburbs and exurbs were built in green fields. Whether natural forested areas, or wetlands, or farmland, an enormous extent of land was paved over to accommodate this type of development, creating enormous swaths of impervious surface and disturbing land patterns that allowed for biodiversity and ecological health (Forman 1995). The impact of sprawl on environmental health is extensive, particularly the impact related to consequences of automobile dependency, which leads to consumption of large amounts of fossil fuels and other petroleum products. The United States is the top consumer of gasoline in the world, with an average consumption of more than 350 gallons per year per capita (Davis 2013). There is a direct correlation between sprawl and the average number of vehicle miles traveled (VMT) per person (Ewing, Pendall, and Chen 2002). Burning fossil fuels produces emissions that are harmful to air and water quality. Air quality can be impacted by vehicular emissions that release carbon dioxide (CO_2), carbon monoxide (CO), nitrogen oxides (NO_x), sulfur oxides (SO_x), volatile organic compounds (VOCs), and lead and metal particulates into the air, increasing mortality and threatening respiratory health (Ewing, Pendall, and Chen 2002; Frumkin, Frank, and Jackson 2004). Additionally, NO_x and VOCs combine to form ozone that affects the planet's ability to regulate its climate (Ewing, Pendall, and Chen 2002; Frumkin, Frank, and Jackson 2004). Comparative studies found that although congestion is the same in outlying and high-density areas, ozone levels are lower in high-density areas (Ewing, Pendall, and Chen 2002).

In addition to emissions, automobile-dependent urban forms require large amounts of impervious surfaces; automobile infrastructure results in diminished groundwater recharge and interruption of the natural nutrient processing of water systems. Stormwater runoff from roads, parking lots, and driveways impacts water quality in a number of ways, including disruption of natural water processing systems and contamination of surface water with pollutants from vehicles (Frumkin, Frank, and Jackson 2004). Increased levels of phosphates and nitrates can cause eutrophication within water bodies, killing many organisms (Gaston, Davies, and Edmondson 2010). Disruption in the water purification and waste treatment cycle can contaminate the water supply, impacting disease monitoring and regulation (Frumkin 2010).

Another unforeseen consequence of this brand of development has turned into a health crisis in the United States and is also being observed in other countries. The obesity epidemic can be directly linked to urban form and mobility habits (Frank and Engelke 2001; Handy et al. 2002; Ewing et al. 2003; Frank, Engelke, and Schmid 2003; Lee and Moudon 2004). The consequences of this epidemic include record incidence of diabetes in both adults and children, cardiovascular disease, and deteriorating mental health related to lack of self-esteem.

Now What: Adapting and Retrofitting

The adaptation of neighborhoods is at once part of the natural evolution of urban areas and a concerted effort to improve quality of life and consequently health outcomes. Some of the anti-sprawl tactics used to stop the momentum of unsustainable development include urban growth boundaries, regional coordination (metropolitan councils), regional tax-base sharing, subsidized housing for low-income households at a regional scale, and regional transit systems (Downs 1999). However, the unforeseen consequences of sprawl offer enough evidence that there needs to be a reversal, and retrofitting of suburbia is, so far, the most viable option (Dunham-Jones and Williamson 2009; Tachieva 2010). It is estimated that urbanites generate one-third of the carbon footprint of suburbanites. Several communities have initiated adaptive reuse of typical suburban structures, such as big box retail and derelict shopping malls; others have started redesigning their layout to allow for more connectivity and to increase mobility options to include walking and biking. Successful examples include the Belmar neighborhood in Lakewood, Colorado; the Cottonwood Mall in Holladay, Utah; and the University Town Center in Hyattsville, Maryland.

A regional response that has greatly improved the way communities deal with this problem is Smart Growth. An initiative that started in Maryland in the late 1990s, Smart Growth is being adopted by cities and regions that need to improve the quality of life of their citizens. Its principles include actions to reverse past sprawl and to prevent future sprawl, as follows (Smart Growth n.d.):

- mix land uses;
- take advantage of compact building design;
- create a range of housing opportunities and choices;
- create walkable neighborhoods;
- foster distinctive, attractive communities with a strong sense of place;
- preserve open space, farmland, natural beauty, and critical environmental areas;
- strengthen and direct development toward existing communities;
- provide a variety of transportation choices;
- make development decisions predictable, fair, and cost effective; and
- encourage community and stakeholder collaboration in development decisions.

The Environmental Protection Agency (EPA) has joined nonprofit organizations and private businesses to form the Smart Growth Network (https://smartgrowth.org) to promote Smart Growth principles and best practices and to disseminate Smart Growth strategies.

Before Smart Growth became the most promulgated sprawl contain-

FIGURE 25.4. São Paulo, Brazil: Parque do Anhangabaú, restored after a highway that had replaced it was rebuilt underground, connects the old and the new sides of this megacity.

ment strategy, Portland, Oregon, had started its own brand of "smart" development based on regional land use planning and strong decision-making agencies (Downs 1999). Since the 1970s Portland has been making decisions and implementing policies that make it, by most accounts, the most sustainable city in the United States today. Portland has been able to adopt and implement several Smart Growth principles, but not all of them. Downs (2005, 369–373) argues that applications of Smart Growth principles face some obstacles: (1) redistributing benefits and costs of development, (2) shifting power and authority from local to regional levels, (3) increasing residential density, (4) raising housing prices, (5) failing to reduce traffic congestion, (6) increasing the "red tape" of new development, (7) restricting profits for owners of outlying land, and (8) replacing "disjointed incrementalism" with regional planning.

Despite the challenges to retrofit suburbia and to implement Smart Growth policies, most communities in the United States realize that they need to adapt and adopt a style of development different from the one that prevailed in the twentieth century. One last obstacle is to get homeowners and "plain citizens" (Downs 2005, 368) to buy into the idea of sustainable neighborhoods. In addition, some policy changes that will enable neighborhoods, towns, and cities to be more sustainable will have to be enacted at the state level (Nelson and Lang 2009). Finally, densification does not work everywhere; re-greening is a good option that allows communities to restore the local ecology and create real estate attractive to buyers in some markets (fig. 25.4). For example, restoring wetlands and daylighting creeks and rivers can create opportunities for waterfront property, which in turn attracts additional investment that can be redirected to declining neighborhoods.

Citizen Science Program: Neighborhoods' Effects on Well-being

The Stanford Prevention Research Center conducted a study in partnership with Place Lab, a San Francisco-based urban design and planning nonprofit, to explore why people feel differently in certain neighborhoods and what factors contribute to comfort or cause stress. Citizen scientists collected data on individuals using a method developed by a Stanford research group. Fourteen local citizen volunteers took the same 20-minute walk in a selected neighborhood and, using a phone app, recorded audio narratives and took pictures to document factors related to the well-being of interviewees. Participants also wore a wrist sensor to record biometric data that are proxies for stress, such as blood pressure. The recorded time- and place-stamped data, which were then transposed to a map, showed how participants' bodies responded to each environment. Environmental factors observed by participants included traffic, noise, safety as a pedestrian, and aesthetic qualities. This small sample showed that any quiet area, not necessarily parks or open green space, offer respite; however, more research is needed to analyze other neighborhood variables. The plan is to repeat the experiment and to include participants from more diverse groups so better conclusions can be drawn from the study (Huber 2018).

Case Study: Via Verde: Healthy Living in the Bronx, NY

Via Verde, the 2013 Rudy Bruner Award for Urban Excellence silver medal winner, is an affordable housing development in the South Bronx that combines environmental restoration with social engagement and equity (Bruner Foundation Inc. 2013). At the time the development was envisioned, the South Bronx was not a very healthy place to live; it was a food desert where asthma and obesity rates were high. The site where the project was to be built was a brownfield site, so remediation prior to development was required (Kimmelman 2011). The developers and architects who partnered to design and build Via Verde decided to incorporate features into the buildings and their surrounding areas that would encourage physical activity, provide natural light and ventilation to all units, and save energy. These features include a fitness center in a prominent location, green roofs with communal garden plots, solar panels, and ceiling fans and windows strategically located to improve ventilation. A LEED Gold certified project, Via Verde comprises 222 affordable housing units, open green space, and commercial space, including a medical facility (Archdaily 2012). It demonstrates how sustainable design can be achieved, even within the budgetary constraints of affordability, when the private sector works creatively and partners with governments to achieve a specific goal.

26

Landscape Aesthetics

Landscape aesthetics is defined as the visual quality or beauty of combined elements, such as flowering plants, water, and trees that cover the land in a defined area. Associated with landscape aesthetics is the concept of an aesthetic experience, described as a feeling of pleasure attributed to certain characteristics of visual quality, such as color, form, and texture that people perceive in landscapes. Aesthetics is described differently for natural areas and urban spaces based on the degree of human influence. Natural scenic beauty, or the "scenic aesthetic," is often described and quantified by elements that indicate ecological quality, such as abundant trees, green ground cover, and clean water, while urban spaces are more often described by the "aesthetics of care" or characteristics that display neatness, order, and stewardship in urban landscapes. For some urban dwellers, neat landscapes indicate good ecological quality, because signs that humans maintain the landscape are often perceived as improving the environmental health (Gobster et al. 2007, 961; Nassauer 1988, 975, 977; Tveit, Ode, and Fry 2006, 238). Visual quality of landscapes in urban areas adds to the image and desirability of a community—important factors in livability, economics, safety, well-being, and social welfare (fig. 26.1).

At first glance the link between landscape aesthetics and urban ecology may not be readily apparent; after all, what does landscape beauty have to do with ecological function in a city? Landscape aesthetics, once considered the domain of designers, is now being embraced by ecologists, conservationists, and land managers who believe it may be the key to saving nature in cities and promoting the use of natural areas to mitigate climate change impacts. The reason for this optimism is that people are willing to pay for and protect what they value, and they place high value on their landscapes for aesthetics and well-being benefits. They are also more likely to engage in pro-environmental maintenance practices when they value nature (Warner, Ali, and Chaudhary 2017, 2). The value or inherent

FIGURE 26.1. Paris, France: a pleasing streetscape with vines on buildings at Au Vieux Paris contributes to the image of a community.

worth of nature is often associated with payment for ecosystem services (PES), including cultural services such as experiencing beauty. Value can be measured by willingness to pay (WTP), or willingness to accept (WTA). These two concepts are used by resource economic experts to determine the maximum price or value a consumer would pay to gain something or avoid the undesirable, and the minimum amount of money a consumer is willing to accept to abandon a good or put up with something negative. Since visual quality is a factor in emotions and beliefs that create positive place attitude (attachment to place or sense of place), aesthetics can play a significant role in WTP and WTA for pro-environmental protections and behaviors. For example, a consumer who finds a high turf landscape attractive is likely willing to pay more for irrigation water and/or would not accept a small tax rebate to abandon a high turf landscape (Nielson-Pincus et al. 2017).

Other areas of land management have recognized the value of aesthetics as a tool to promote resource protection and conservation goals. Social and environmental scientists, conservation and restoration experts, land planners, geographers, and land managers all consider aesthetic perceptions in the public's acceptance of proposed projects. Promoting the beauty of natural landscapes as a basis for conservation is grounded in the knowledge that the care and stewardship of landscapes often depend on aesthetic appeal; studies have shown that people are more likely to care about and take better care of beautiful landscapes because of their emotional response to the aesthetic experience. This sentiment is also important for participation in citizen science programs. The motivation to participate as an environmental volunteer is often based on desire to learn, desire to help nature/improve the environment, enjoyment of looking at nature (or pictures of nature), enjoyment of the beauty of nature, and personal environmental values (Bruyere and Rappe 2007, 510; Raddick et al. 2010).

What: Characteristics of Landscape Aesthetics

Aesthetics is a combination of qualities, such as shape, color, and form that pleases our senses, especially the sense of sight. Humans gain approximately 65% to 85% of their information about the environment through visual perception (Medina 2018), and we typically describe the environment through visual attributes, although the total aesthetic experience includes smell, touch, hearing, and taste. Two different approaches are used to describe aesthetics in landscapes: one is the use of formal design elements and principles commonly used by artists and designers, and the other is a set of concepts describing conditions or indicators used to analyze the visual character of landscapes.

When selecting and organizing plants in a garden, the designer uses design elements such as color, form, texture, and size to create an attractive view (Hansen 2010, 4–6). People often describe their preferred plant qualities using these same terms, noting a fondness for large, brightly colored flowers, wide leaf width, and green color foliage with texture (Kendal, Williams, and Williams 2012, 35–38). Traditional design principles, including balance (equal visual weight), unity (harmony of parts), repetition (elements that create pattern), dominance (clear focal points), and complexity (variety of elements) are used to create interest in landscapes (Hansen 2010, 7–9). These same principles are also used by nonexperts to describe general preferences for organized human-created landscapes, including a clear focal point, smooth even ground cover, expansive views, and evidence of human influence (Balling and Falk 1982, 9; Kovacs et al. 2006, 61).

Designers and planners are interested in certain conditions or indicators that illustrate how people experience, perceive, and react to landscapes to help them evaluate land use policy and development activities.

FIGURE 26.2. Copenhagen, Denmark: visual character of a landscape is important to perception and experience. People react positively to a landscape view that includes water and historic architecture of a church.

Concepts used to describe conditions include stewardship, coherence, and disturbance. Coherence and stewardship denote a sense of order and care through repeated use of patterns, color, and texture to create harmony, while disturbance is the lack of coherence, usually caused by views that deviate from context. Other descriptive indicators include historicity, the visible evidence of time, and imageability, a strong, memorable visual image that is usually enhanced by historic features (fig. 26.2). Visual scale and complexity describe both the openness of a view and the diversity and richness of features within that view. Naturalness and ephemera are related indicators, one being close to a natural state and the other a sense of seasonal change, which indicates nature (Tveit, Ode, and Fry 2006, 238–246). These concepts are important to determine actions and policies that enhance visual quality and preserve the character of a landscape. Policies can also mitigate loss of aesthetics due to disturbance, prevent loss of historical richness or heritage, maintain views, preserve special features and diversity, and maintain natural features important for the aesthetic experience that enhances human well-being.

While studies clearly show that people have landscape preferences, and the concepts and techniques for creating and analyzing aesthetic character are well established, no central theory exists that explains why people have certain landscape preferences. Currently there are two broad landscape aesthetics paradigms; one takes an objective approach, and the other a subjective approach. The objective approach uses the formal elements and principles of design to measure the inherent aesthetics of landscapes. Although widely accepted by design professionals, it is only somewhat useful for explaining the preferences of nonprofessional viewers. The subjective approach contends that preferences are not based on formal design attributes, but on people's understanding of the landscape and the perceived benefits such as human well-being. Theories that support the notion that landscape preference is important for human well-being and survival rationalize the subjective paradigm, including the evolutionary theories and the cultural preference theories (Maulan, Shariff, and Miller 2006, 25, 30–31). Evolutionary theories are based on an innate preference for landscapes that support survival. The most prominent is the prospect/refuge theory, which suggests the idea that humans historically preferred savanna landscapes where we could have both prospect, the opportunity to see, and refuge or protection, the ability to see without being seen. From the evolutionary perspective this may help explain our strong (and somewhat curious) preference for turf and shade trees (fig. 26.3). Cultural preference theories are based on our cultural background and personal attributes, where the common landscapes of our community and our childhood experiences in nature form our preferences. Our environmental values and beliefs can also influence how we perceive landscapes. If we believe that plants and nature are critical to our survival and have great value to us personally, we are more likely to protect and care for landscapes and are

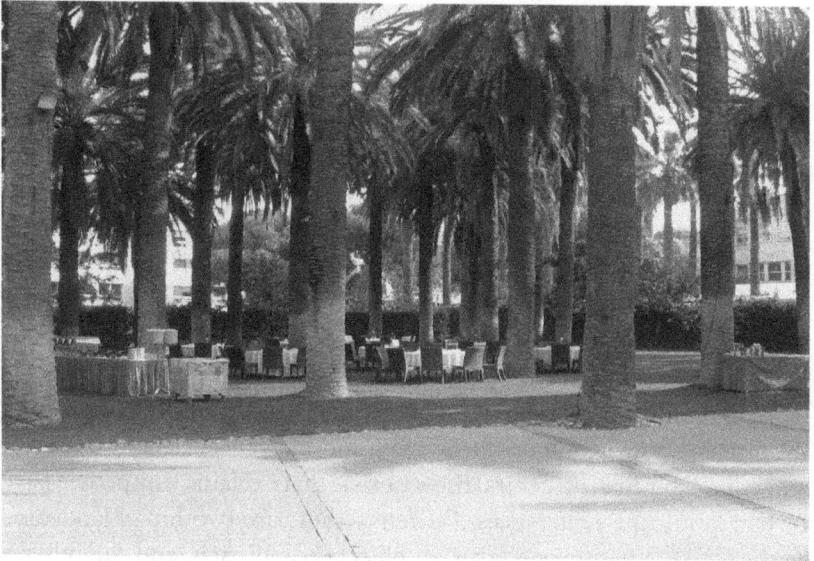

FIGURE 26.3. Izmir, Turkey: a garden scene with large trees and turf is a cultural preference that provides protection yet allows visibility.

usually more accepting of "natural-looking" landscapes. Most researchers believe a synthesis between the evolutionary and cultural theories is the most appropriate for urban landscapes where social norms are highly influential (Tveit, Ode, and Fry 2006, 231–232).

Although numerous studies have explored what people prefer in landscapes and why, there is less knowledge about the role of social influences. The concept of social norms (expected behaviors) have been shown to greatly influence our preferences, especially the desire to be good neighbors and "fit in" with our neighborhood, which often compels us to mimic the landscapes of our neighbors' yards. In fact, the preferred landscape in most neighborhoods is best predicted by what residents believe their neighbors' preferences are. However, perceptions about neighbors' preferences often prove to be false; for example, one study found that homeowners preferred a 50% native plant garden, but mistakenly believed that their neighbors preferred a 0% native plant garden, which influenced their attempts to "fit in" by using fewer native plants (Peterson et al. 2012, 10–11). While natural beauty is more often associated with expansive natural scenes, the scale at which most humans regularly experience landscape aesthetics is referred to as the human perceptible realm, the space where we perceive and interact with urban nature. In this realm people often mistakenly relate pleasurable aesthetic experiences with good ecological quality, but aesthetic preferences and ecological goals don't often align (Gobster et al. 2007, 960, 962). However, using the concept of "aesthetics of care" can be a means for promoting healthier ecosystems in cities. Aesthetics of care can demonstrate alternative forms of care, for example by using native or naturalized plants in traditional settings, using small patches of mown lawns—just enough to show a familiar label of care, and using horticultural maintenance techniques on native plants such as pruning. The challenge is to plan, design, and manage landscapes that link aesthetics to ecological goals (Gobster et al. 2007, 967, 970).

So What: Environmental Impacts of Aesthetic Landscape Choices

Aesthetics can have two opposing effects on the environment: on one hand, the desire to conform to social norms for visually appealing landscapes can have negative consequences, and on the other hand the emotional appeal of beautiful landscapes can motivate people to use pro-environmental behaviors when caring for landscapes. Landscape preferences often drive plant selection and maintenance behaviors, setting up a potential conflict between the desire for visual aesthetics and ecologically sustainable landscapes. Social norms can also influence how we maintain our yards. People seek advice on how to maintain a landscape primarily from their neighbors, especially those with neat and visually appealing landscapes. This can lead to undesirable and mostly unintended environmental consequences created by the collective maintenance practices in neighborhoods with similar landscapes. The desire to fit in with neighbors with a well-kept and appealing yard can escalate the use of maintenance products, contributing to environmental degradation. The quest for the perfect lawn is not always voluntary; many communities have homeowner association (HOA) landscape covenants that impose maintenance requirements and expectations for visual quality. A typical code may require a certain percentage of turf and plant beds in the yard and a mandate for weed-free green lawns all year. With this expectation, homeowners adopt yard care routines that often overuse water, fertilizer, pesticides, and herbicides (Peterson et al. 2012, 2). Social norms are also driven by lawn product development and marketing. Advertising strategies targeting do-it-yourself homeowners, who make up about half of consumers, set the bar high for aesthetic quality of landscapes by using subtle cultural and social normalization for an ideal lawn. Homeowners use ten times more fertilizer and pesticides on lawns than farmers use on crops, which contributes to water quality problems and affects biodiversity (Polycarpou 2010). Insect numbers and variety decrease with increased pesticide use, and overuse of herbicides decreases plant variety, limiting food sources and shelter for birds and other urban wildlife.

Although it is apparent that our landscape choices to create visually appealing lawns can have negative impacts on the environment, persuading people to change is challenging. Many barriers to pro-environmental behavior change exist, including knowledge of issues, environmental values, emotions, priorities, social expectations, and financial restraints. Educational programs and materials are available from universities through extension programs and through government agencies, but gaining knowledge does not often translate to behavior change. In one study, participants initially rated ornamental landscapes with turf and beds more aesthetically appealing than more natural forest-garden landscapes. After several workshops on the design, management, and sustainability of the landscapes, the participants perceived the ornamental landscape to be even more pleasing than they had initially, but also the least sustainable, while the natural landscape was viewed as most sustainable and least appealing both before and after the workshops, demonstrating that aesthetic appeal and sustainability may be hard to reconcile in landscapes (Beck, Heimlich, and Quigley 2002, 167–169).

Now What: A New Ecological Aesthetic

Landscape aesthetics have recently taken on a different meaning for designers and planners with a new theoretical framework for "ecological aesthetics," the design of ecologically sound landscapes that both conserve biological diversity and ensure care and stewardship through aesthetic appeal. For example, three different aesthetics-based approaches are currently being used to encourage more environmental lawn behaviors, including encouraging people to reduce (but not eliminate) turf in their yards, using social norms to increase acceptance of a less-than-perfect lawn, and using science to develop new turf cultivars that are water conserving and stay green year-round without fertilizer.

Forward-thinking HOA covenants have management plans that include new strategies for more environmentally oriented maintenance, including integrated pest management (IPM) to control insects with biological controls. Developers are using new irrigation technologies such as soil moisture sensors and drip irrigation to conserve water, and plant breeders are developing plants that are pest resistant, low water users, and hardy in temperature extremes. Developers are also recognizing the value of aesthetics for creating desirable communities and are writing new development codes, such as form-based codes that regulate the look of landscapes and buildings, with the goal of harmonizing building and landscape design to achieve a cohesive, visually pleasing community.

New technology is also being used to learn more about how people perceive landscapes and form preferences. Eye-tracking technology has been used by study participants to view landscapes with varying amounts of turf and plant material. Special glasses record the time of gaze and eye movement to determine what parts of the landscape they prefer and are most familiar with. Using gaze time as an indicator of preference is based on studies that show people tend to look longer at things they like and are familiar with (Dupont, Antrop, and Van Eetvelde 2014). Education programs are also benefiting from technology. More information is available online in various forms, including videos, smartphone applications, and interactive programs, to help homeowners make pro-environment design decisions. Smartphone applications, such as FlowerChecker, NatureGate, PlantSnapp, and Leafsnap, help people identify plants by simply taking photographs, and plant databases guide consumers through design decisions to select the best plants based on certain criteria. Programs can estimate the amount of water a landscape will use based on size and plant material, and irrigation apps can help homeowners with smart systems water only when needed.

Citizen Science Program: Globe at Night

The National Optical Astronomy Observatory (Arizona State University and SciStarter) developed the Globe at Night project to raise awareness of light pollution after they learned that 7 out of 10 people in the United States had never seen the beauty of the Milky Way Galaxy in the night sky because of light pollution. The concern about light pollution is not only that it affects our view of the stars, but also that it wastes energy, causes sleep disorders, and disrupts the sleeping and breeding habits of animals. During a 10-day campaign, the participants go outside before 10 pm each night to a location where they are not under or near a light and count how many stars they see in a constellation. They compare what they counted by looking at stellar charts and selecting the one that most closely resembles what they saw. After marking their location, they note whether it is cloudy and submit the data. Worldwide observations are produced immediately, so participants can compare light pollution levels at their location against other locations around the world. Educators and astronomers hope that participants will realize the night sky is a natural resource worth protecting and advocate for regulations to reduce light pollution (Discover 2019).

Expert Insight: Fremantle, Australia

Fremantle is a district of the city of Perth in Western Australia, an area known for a dry climate and frequent water shortages. Although most residents in Perth have green lawns and introduced plants in spite of the water shortages, most residents of Fremantle have adopted native gardens with no turf. Curious about this condition, researchers at Curtin University in Perth interviewed 12 residents to learn more about what prompted the homeowners to use native gardens. Primarily, participants agreed that the economic benefits of saving water were valued above environmental and aesthetic concerns, even describing buying plants for their water-saving qualities. They also mentioned a dominant cultural value system in Fremantle (it was Australia's first carbon-neutral local government) that made it a safe setting to adopt pro-environment behaviors. Many commented on being inspired by their neighbors and the flow-on effect as more and more neighbors converted their yards to natives to meet the neighborhood norm. Many claimed a sense of ecological responsibility and a desire to be good global citizens, lamenting that these values were not a priority of others in Perth. However, participants also expressed a subtle desire to have more European-style gardens that aligned more with their preferences for aesthetics, saying that their garden looked best in spring when everything was colorful, green, and lush. And while they wanted to use native plants, they also wanted a controlled, neat appearance, complaining that the natives didn't always look good. They also mentioned that when buying plants, they looked for natives that were lush, colorful, and had a long flowering period, and some would even sneak in an exotic or two to provide more color. There also appeared to be a subconscious belief that the color green was a symbol of nature and a sustainable environment, with one participant saying he had used fake grass in the backyard, justifying his behavior by saying that he only wanted it to look healthier and cooler in summer, and that surely the insects could thrive because there were only a few small squares of grass. The competing preferences and behaviors illustrate a tension between a desire for aesthetics, the desire to be a good environmental citizen, and the desire to follow the social norms of the community, with some even admitting that they took care to hide certain unsustainable behaviors from their neighbors (Uren, Dzidic, and Bishop 2015, 77, 79, 80–82).

27

Residential Landscapes

Residential landscapes are privately owned green spaces in urban/suburban areas that play a unique and significant role in the urban ecology of a city. About 40% of the land cover in urban areas is used for homes and yards, with one-quarter of that land being residential outdoor space. The average U.S. residential lot is one-third acre, with about 60% lawn, 20% plants or hardscape, and 20% house and driveway (Cook, Hall, and Larson 2012; Whiteman 2012). Residential yards impact urban ecology in many ways; most notably, they provide desirable outdoor amenities and contribute to the green infrastructure of urban areas. Although they can have negative environmental impacts, they also have great potential to be important biodiverse habitats for wildlife and vegetated areas for climate change mitigation.

An important concept related to homeowner yards is the fundamental notion of private property rights that make restrictions and policies for environmentally friendly practices difficult to apply and enforce. For example, despite efforts to encourage reduction in the use of pesticides and herbicides in yards, the use of chemicals is difficult to regulate because of private property rights (Robbins and Birkenholtz 2003, 183, 191). The task of changing landscape behaviors is left largely to social marketing campaigns to encourage management practices such as integrated pest management (IPM) or reduced fertilizer or water use. Studies show that, overall, residents' environmental values and attitudes have limited influence on stewardship and maintenance practices because of larger institutional forces, such as HOA rules, government ordinances, and social norms at the household and neighborhood scale. Notable differences between front and backyards illustrate this condition; generally backyards (the private realm) reflect the homeowner's landscape preferences and desires, whereas front yards (the public realm) reflect restrictions and neighborhood social norms (Cook, Hall, and Larson 2012).

Many people have an interest in residential landscapes, including green industry professionals, developers, gardeners, and, more recently, urban ecologists, who are just beginning to study the contribution of neighborhood yards to urban ecology. The 2018 National Gardening Survey reported that American gardeners are spending a total of about $52 billion on all lawn and garden retail sales, with an average annual household spend of $503 (Garden Research 2018). The number of households purchasing landscape design, installation, and maintenance services has doubled in the past six years, led by greater interest on the part of the 18–34-year-old age group in millennial households (Garden Research 2018).

What: Physical Characteristics and Social Value of Residential Landscapes

Urban residential landscapes are mostly novel human-made landscapes that represent a wide range of ecological habitats. While green lawns with shade trees are perceived as the ubiquitous, homogeneous American yard, landscapes actually vary considerably between neighborhoods. Although the biophysical properties can vary greatly, predictable patterns are observed based on several factors, including yard size, housing density and age, lifestyle behaviors and socioeconomic status of the homeowner, and legacies of former land use. For example, gardens in older suburbs tend to support more wildlife, especially birds, with mature trees and shrubs, while newer suburbs tend to have a greater proportion of turf and smaller trees with less total plant biomass than older suburbs (Cook, Hall, and Larson 2012; Douglas 2011, 264, 265). A review of studies of residential yards shows certain patterns for plant community composition; for example, the dominant flora are non-native species, but there is overall high diversity of individual species, and species richness is often positively related to yard size. Plant choices, garden styles, and species diversity across neighborhoods are all positively related to education and socioeconomics. Proximity is a key factor in shared yard characteristics among neighbors; adjacent yards tend to repeat plant combinations (the flow-on effect) throughout the neighborhood. In addition, proximity to natural habitats and corridors also affects abundance and variety of animal and insect species in yards. Most studies have focused on the front yard because of ease of field observations, but there are differences in front and backyard plant structure. Front yards tend to be more highly maintained (fig. 27.1), likely because of social pressures and public visibility, while backyards tend to have more

FIGURE 27.1. Gainesville, Florida: a nicely maintained front yard represents a typically preferred residential landscape.

flowers and wildlife resources with greater vegetation complexity and more native plants to attract birds and pollinators (Belaire, Westphal, and Minor 2016, 401; Cook, Hall, and Larson 2012). While soil characteristics are similar to those of other developed urban areas, older residential soils and soils from yards that were former agricultural fields (legacy land cover) contain larger total pools of organic carbon and nitrogen than new properties. Some studies show that landscape legacies, such as previous land cover, neighborhood lifestyle, and socioeconomic characteristics, are good predictors of current land cover and continue to shape cultural norms for yards over time. For example, older established neighborhoods with high-value homes tend to retain the traditional landscapes with turf and flower beds (fig. 27.2) that have been in neighborhood yards for decades (Cook, Hall, and Larson 2012).

Although the physical characteristics of yards vary widely, there is more consensus on the social significance of yards and gardens. Homeowners value their yards for several reasons, including for beauty and relaxation, a comfortable place to entertain, and a safe place for children to play and pets to exercise. In a ranking of the benefits of yards, aesthetics and relaxation ranked first, and environmental benefits such as cleaning and cooling the air and providing a natural water filter rank last (AgEdLibrary n.d.;

National Gardening Association Survey 1999). Homeowners also believe a well-maintained yard reflects positively on the owner (pride of ownership), helps beautify the neighborhood (pride in community) and increases real estate market value. Many homeowners prefer master planned communities that have uniform aesthetics and well-manicured landscapes as a sign of social distinction (Cook, Hall, and Larson 2012).

Gardeners tend to have personal relationships with their yards, often describing them as a place where they can connect and interact with nature, especially with the wildlife in the garden. Gardens are also seen as a place where their own creativity and identity can be nurtured and displayed, and where they can control the sometimes-perceived messiness of nature. Gardeners often talk about "taking care" of the environment and having a kinship with nature. When participants in one study were asked to photograph things that mattered to them in their garden, the most photographed features were exotic and unusual plants, followed by structures and views. In talking about their garden, they described their gardens as giving them sense of purpose and self-worth and a way to improve their unattractive neighborhood. Homeowners sometimes describe their maintenance activities as a "duty of care" to be responsible for the environment, to help nature and fix what we destroyed, and to support the "critters" and

FIGURE 27.2. Perth, Australia: large flowering plants in a corner garden are more typical of older residential yards.

bugs by planting native plants. This suggests that studying a socioecological view of neighborhood yards and the value they provide is important for social marketing messages aimed at encouraging more pro-environmental behaviors (Freeman et al. 2012, 137, 139–142).

So What: The Impact of Residential Landscapes on Urban Ecology

Several environmental problems are associated with residential yards that are not much different from the problems in other urban green spaces: turf monocultures and the use of potentially invasive plants that escape to natural areas. Problems more closely associated with residential yards include pet waste and loss of wildlife, particularly birds lost to predation by domestic cats. Monocultures and fragmentation of landscapes by lawns also impact the reproduction and dispersal of many wildlife species, especially birds (Robbins and Birkenholtz 2003, 181, 184).

Large amounts of turfgrass in suburban yards is the most problematic feature because of the maintenance required for the desired look, but it is also the most difficult landscape choice to influence. Many underlying social and institutional dynamics support the homeowner's choice of turf, including neighborhood social norms, HOA policies and regulations, and the homeowner's own perceptions and preference for a well-maintained lawn (Dorsey 2010, 96). An interesting paradox about lawn care is the fact that well-educated, high-income homeowners who belong to environmental groups or have "pro-environment" attitudes are the most likely to use increased amounts of chemicals for lawn maintenance. Studies estimate from 50% to 84% of U.S. households apply fertilizer and 64% to 75% apply chemical pesticides (Cook, Hall, and Larson 2012). This usage is partly attributed to the homeowner's ability to afford the treatments, but also to the perception that activities that reduce pests and make yards look better are seen as safeguards that improve the environment (Dorsey 2010, 96; Helfand et al. 2006, 237). One reason homeowners have not embraced alternative ground covers with lower maintenance requirements is the perceived lack of aesthetics and a perception of greater expense and time to maintain (Robbins and Birkenholtz 2003, 183, 190).

Although the perception of lawns is primarily negative, turf can provide environmental benefits with proper maintenance practices, including fewer chemical inputs. Lawns sequester carbon, mitigate the urban heat island effect, protect and improve the soil, provide habitat for insects and soil microbes, and improve groundwater recharge by filtration (Blaine et al. 2012, 258). Management practices vary widely based on such things as household income, the homeowner's decision to hire professional service, and the owner's knowledge of different management practices. Owners' personal aesthetic preferences and ideas and the type of residential environment (suburban, urban, or rural) also influence management decisions. Often homeowners will be influenced by their neighbors' use of chemicals and what they believe are the environmental impacts of the chemicals (Cook, Hall, and Larson 2012). Unfortunately, homeowners often rely on biased sources such as relatives and friends, sales marketing, and retailers for information on turf management. Their own perceptions also make it difficult to influence lawn care behavior; for example, a neighbor's use of a lawn care company is perceived to have a negative impact on water quality, but a positive impact on property values and neighborhood pride. However, there is also a perception that when they, or their neighbors, apply their own chemicals, water quality is not affected (Blaine et al. 2012, 257, 261).

Now What: Strategies to Improve Residential Landscapes

Several unique strategies are being employed to improve the environmental benefits of residential yards. Connecting landscape aesthetics with ecological function has created a new paradigm that equates function with beauty. Linking landscape aesthetics with ecological function is a challenging concept but may be useful for persuading homeowners that beauty in nature is "more than skin deep." Most homeowners judge the ecological value of a landscape on its aesthetic appeal, viewing green grass, flowering plants, and lack of weeds as a healthy landscape due to active and careful management. Without the knowledge and expertise needed to identify ecological indicators of health, the average homeowner judges on visual qualities presumed to indicate good health. To remedy this misperception, a relatively new area of research is exploring the ability of visual indicators to inform us about ecological function. Aesthetics and ecological function share several indicators, and the concept is to link aesthetic preferences to ecological functions in order to increase ecological function through aesthetics. Design concepts used to describe visual aspects of landscapes such as disturbance, complexity, and imageability can all be linked to ecological function. For example, most people can visually recognize disturbance due to very different plant species and a landscape that does not "fit" with the surrounding area. This can also be a visual cue for loss of function

FIGURE 27.3. Gainesville, Florida: a variety of shrubs, trees, rocks, and turf show a complex, but orderly landscape.

due to loss of natural vegetation. Complexity refers to diversity and richness of elements, and it is a highly visual component of landscapes that is also an indicator of biodiversity. Most homeowners appreciate a degree of complexity for a more interesting landscape, creating the opportunity to make landscapes more visually appealing with a variety of plants that also increase diversity and habitat (fig. 27.3). Finally, imageability, which is created by visually pleasing features that give a landscape identity and sense of place, is often due to the same features that create unique biodiversity in the landscape. For example, plant communities found only in certain locations can be used in residential landscapes to enhance the imageability of an area (Fry et al. 2009, 936–942).

To promote better urban ecological practices on residential property they must be compatible with needs and interests of homeowners rather than an either-or choice. The YardWorks project (see Case Study for a description of the process), developed eight urban ecological design strategies to meet the needs of a homeowner and enhance the ecological conditions of residential neighborhoods. Each strategy includes the rationale and methods for the design, and the performance metrics for measuring the benefits. The first step is to study the surrounding landscape network for opportunities to enhance connectivity with stepping-stones (vegetated patches) and green corridors. For example, strategically planting trees to connect patches supports the design concept of enclosure to create outdoor "rooms." The second strategy is to improve structural deficiencies of plant material by increasing vertical plant layers using nearby natural habitats as a guide. This strategy matches the basic design concept of layering short, medium, and tall plants for interest and depth perception. Strategy three, creating refuge habitat for birds by clustering shrubs, also improves the visual quality and legibility by use of the design technique of massing plant material. Strategy four recommends using a plant palette that provides forage for birds year-round by selecting plants with seeds, berries, or nuts that also match the form, size, and texture characteristics of plants desired by homeowners. Strategy five recommends providing for pollinators by using a variety of plants with overlapping bloom times in clusters of the same species. Groupings more than three feet in diameter are more attractive to pollinators and also work well in more traditionally designed flower beds. Increasing plant diversity (strategy six) provides species with habitat and creates visual interest in the landscape. Site features such as rock walls and boulder and brush piles also provide habitat and can be used to complement the design concept with thoughtful placement. A final ecological strategy recommends using water features or rain gardens and vegetated swales for habitat, stormwater control, and as decorative design features in the landscape (Cerra 2015).

Little attention has been paid to residential yards as potential green spaces for improving community ecosystems. Barriers to access and study of private property mean that little information about the ecological function of typical residential landscapes is available. Citizen science may offer a way to bypass these barriers to collecting data and create a new way to advance conservation in residential yards. Citizen science models rely on individuals to gather data at many different locations to build large databases accessible to scientists for study. The configuration of typical neighborhoods fits the CS model, with individual homeowners collecting data in their yards to crowd-source with other homeowner data. The process uses the concept of a new adaptive citizen science approach in which scientifically informed management strategies are applied, evaluated, and revised to improve outcomes. One example would be managing yards to reduce mortality of native and migratory birds, such as controlling domestic pets and improving the wildlife habitat of the neighborhood. Adaptive management processes would involve collecting data on bird populations and manipulating habitats with a series of treatments that are measured and used to develop new strategies. This would involve recruiting residents and coordinating their data collection efforts so that ultimately the cumulative impacts of many independent land-use decisions by homeowners could be coordinated to create a major large-scale outcome for the entire neighborhood and community (Cooper et al. 2007).

Citizen Science Project: YardMap

YardMap is a citizen science project that invites volunteers to share their sustainable yard practices with others to support people who want to manage landscapes for wildlife. Participants generate a birds-eye view of their property by locating it on a Google Map interface using YardMap's simple point-and-click mapping tools. The project is the first one to collect spatial map data from the public on a large scale. Participants can share their maps, showing ecological details such as native plants or bird feeders, and their strategies and success with other YardMap volunteers and scientists at the Cornell Lab of Ornithology, creating a full social network within a citizen science project. Lab scientists are using the information to address large-scale impacts of residential yards on birds and the project (Whiteman 2012).

Case Study: YardWorks

YardWorks was a three-year studio research project at Cornell University in collaboration with the Cornell Laboratory of Ornithology's YardMap program and Cornell Cooperative Extension to investigate how urban ecological design interventions could best fit in residential settings. The goal of the studio research was to develop urban ecological design strategies that are compatible with aesthetic and program goals of homeowners. The project began with a collaborative visioning process to set stewardship goals at the neighborhood level with community members, which resulted in a set of site-specific design strategies to meet the goals. Strategies were tested by working with 35 individual homeowners to incorporate the strategies into site designs acceptable to the homeowners. As a result, a series of urban ecological design strategies were developed to improve landscape performance on private properties with potential metrics for measuring the benefits. Each studio project resulted in a consensus-based vision, a set of stewardship goals based on the vision, an analysis of existing human and natural systems for the neighborhoods, a series of design strategies, and a set of proposed designs for the properties of the study participants. One goal of the project was to give landscape architects a better understanding of ecological design interventions for residential settings, including "what to do" and "how to do it" strategies, but the strategies can also be used by homeowners (Cerra 2015).

28

Urban Ecology

Knowledge, Education, and Professions

Cities are complicated. Many social and environmental dynamics play out daily in cities that require two types of knowledge: science-based (expert) knowledge and culture-based (lay) knowledge. Expert knowledge usually refers to knowledge acquired through formal training, and lay knowledge, sometimes referred to as soft knowledge, is acquired through cultural and social experience. Both are equally important in the realm of urban ecology and should be used in all planning, design, policy, and decision-making processes. Expert science-based knowledge can be found in several disciplines that usually track in four broad domains: (1) the humanities, the study of human phenomena; (2) the social sciences, the study of social phenomena; (3) the natural sciences, the study of natural phenomena; and (4) the applied sciences, the application of scientific knowledge to technology and inventions. All four are related to some aspect of urban ecology.

The humanities and social sciences study human societies and cultures and comprise several disciplines. Fields of study often include four thematic groups: (1) literature, ancient and modern languages, performing and visual arts, and music; (2) philosophy, religious studies, economics, law, and politics; (3) anthropology, archaeology, and geography (physical and human); and (4) psychology and environmental sociology. Every theme in the humanities has a direct or indirect connection to the study of nature in cities.

Natural sciences is the study of how the world and the universe work and includes five major branches: chemistry, astronomy, physics, earth science, and biological and life sciences. Of these, the earth sciences and biological and life sciences are most relevant to urban areas. More recently, there has been an effort to better integrate traditionally discrete disciplines and to link educational objectives to practical action (Friedmann 1987; Moore 2001). Environmental studies is an example of integration of social,

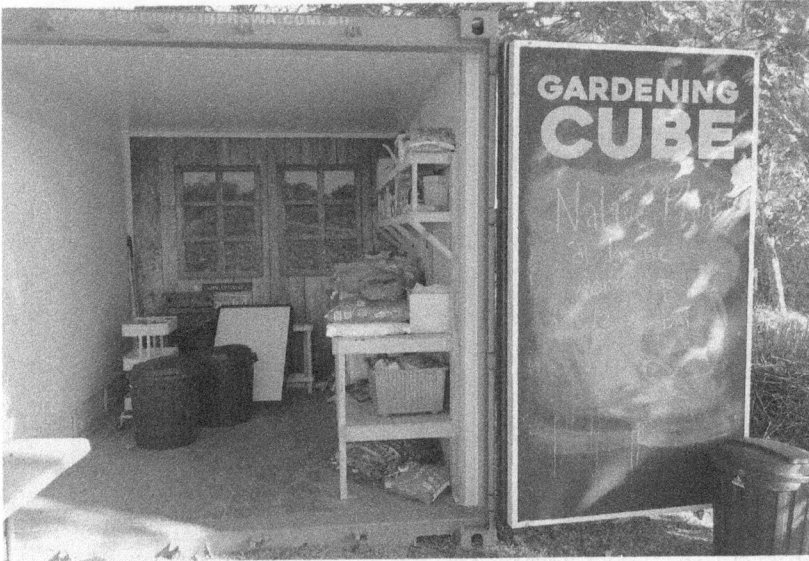

FIGURE 28.1. Perth, Australia: a gardening "cube" on Curtin University campus features native plant workshops for all students on campus.

humanistic, and natural science perspectives, studying not only the relationships between humans and the natural environment, but also how chemistry, geology, and biology relate to human activities (fig. 28.1). As a result, we now have a better framework for studying the impact of human activities and interventions in the natural environment on biodiversity and sustainability (Study.com 2020).

The applied sciences include the broad fields of medicine and engineering and related disciplines that apply to urban ecology, such as architecture, landscape architecture, building construction, and urban planning. Medical sciences are typically considered a separate discipline from natural sciences, but they should be considered in urban ecology because subdisciplines such as public health and epidemiology are connected to urban ecology through the health and well-being benefits of the environment. Civil and environmental engineering are directly connected to the environment through the construction and maintenance of natural and built projects. Architecture and landscape architecture have a more direct impact on environmental conditions; these are the disciplines that today are specializing in green buildings, low-impact development, and other strategies to incorporate development into nature and to preserve the environment. Likewise, building construction professionals are specializing in green buildings and the various technologies being developed to make buildings more energy efficient and to reduce the impact of the built environment on the natural environment. Finally, urban planning has a significant role to play in urban ecology. Virtually every urban planning subdiscipline studies factors related to the impact of development on the environment: transportation relates to infrastructure and pollution; housing and community development relates to density and buildings; urban design relates to urban form and sprawl; and environmental planning relates to the conservation and integration of the natural and built environments.

What: Professions Related to Urban Ecology

Urban ecology is a multidisciplinary, multi-expert area of study and practice that requires specialized knowledge from experts and local knowledge from citizens to help cities realize their full potential. Areas of professional expertise needed in cities include planning, design, and construction; ecology and natural systems; and politics, law, and economics.

Planning, design, and construction professionals who work in the field of urban ecology include urban planners, environmental planners, landscape architects, architects, civil engineers, and environmental engineers, among others. These professions work in tandem to design, plan, and build the structures and green spaces of the city. Urban planners have knowledge about land use policy, community development, urban design, transportation, and infrastructure; examples of projects include master plans, transit network plans, subdivisions, and design of public spaces (Friedmann 1987; Alexander 2005). Environmental planners are urban planners who focus on ecology and sustainability; they may work as consultants or government contractors who strive to minimize the environmental impact of projects while making sure planning and environmental regulations are followed. Landscape architects specialize in design of the natural built environment; examples of projects include development plans for community neighborhoods, urban parks and gardens, city plazas, and residential gardens. Architects plan and design buildings; they have recently become more involved in ecology with the growing use of biophilic elements as part of building design. In addition, there are buildings that impact the surrounding environment; thus architects need to consider how their buildings can be ecologically integrated into environmental systems. Civil engineers deal with the design and construction of buildings and urban infrastructure projects, including roads, canals, dams, stormwater systems, and bridges. Civil engineers have become key players on urban teams as many cities are dealing with aging infrastructure and need to repurpose facilities and build new sustainably designed structures. Environmental engineers solve environmental problems; they may work with recycling, waste disposal, water and sewage infrastructure, and water and air pollution control, among other issues related to environmental sustainability (Department of Labor 2018).

Ecology and natural systems professionals include urban ecologists, arborists, urban foresters, land managers, and natural resource managers who typically work in the field and are responsible for the observation, care, and management of natural resources. Urban ecologists apply ecological principles to the development of urban areas to create more sustainable ecosystems and resilient cities. Arborists and urban foresters keep urban forests healthy and safe for citizens. Land and resource managers are closely aligned. Urban land managers work in property, real estate, and community associations to make sure property is well maintained and operations run smoothly; they may also manage green spaces, such as parks and wildlife corridors, and are often responsible for legal issues such as property rights. Natural resource or conservation land managers usually work with land trusts or conservation organizations to protect and preserve habitat. They implement conservation and sustainability plans for parks, nature preserves, and historic sites. Many cities and counties have conservation or natural resource land managers who work with city planning staff to acquire and protect green spaces for both preservation and recreational purposes. Conservation land managers are starting to play a more integral role as cities seek to expand their green spaces (Study.com 2019).

Politics, law, and economics include professionals who specialize in connecting environmental, legal, and economic issues to policy. Positions include city managers, environmental lawyers, environmental or resource economists, and experts in environmental politics and policy. City managers have degrees in public administration and are responsible for managing the city budget, implementing policies adopted by city councils, and advising council members on the best courses of action. In this regard they are instrumental in getting environmental policies adopted and funded (Villanova University 2019). Environmental lawyers represent clients in legal issues related to management of land, land titles, management of water resources, species protection, wetland protection, waste and other hazardous materials, and air quality, among many other environmental issues. Areas of legal action include ecology, sustainability, and stewardship; lawyers who can advise clients on green standards and sustainability are in demand (Environmental Science 2019a). Environmental economists study the economics of natural resources, including extraction and use, and waste products returned to the environment. They conduct cost-benefit analyses of industrial activities or proposed regulations and develop sustainable policy recommendations (Environmental Science 2019b). Experts in environmental politics and policy develop environmental policy and also work as political analysts. Their work focuses on issues such as climate change, pollution, and energy (Department of Political Science 2019).

Social and health professionals working with urban ecology include environmental sociologists, mental health experts, public health experts, and epidemiologists. All are focused on the benefits the environment contributes to human health and well-being. Environmental sociology is a field of sociology that studies the interactions between societies and the environment, and societal issues such as environmental injustice. Most environmental sociologists work in research and academics (Environmental Science 2019c). With the recent pandemic, mental health experts, public health experts, and epidemiologists have become inextricably involved in the solution to problems exacerbated by urban lifestyles. In addition to professionals, academics may contribute knowledge and expertise to all the fields mentioned above, acting as consultants to city governments and planning offices.

So What: Licensure and Certifications for Professionals

Many professional positions require degrees from accredited university programs, and/or licensure or specialized training certificates. Degrees and licensure are proof that a person has the knowledge and skills to apply to problem-solving and to implementation of solutions in their area of expertise. Professional licenses and certifications are designed to regulate the professions and to ensure the health, safety, and welfare of the general public. Examples of certifying and accrediting bodies in the United States include the American Society of Landscape Architects (ASLA) and the American Planning Association (APA), which are the professional associations for landscape architecture and planning professionals, and the American Society of Civil Engineers (ASCE) and the American Institute of Architects (AIA), which are the professional associations for engineers and architects.

In addition to certified and registered professionals with full responsibility for projects, technicians working in these professions may pursue certificates that allow them to perform specific installation or maintenance work in these areas. These certificates and special licenses may be issued by trades associations and also by governments seeking to regulate specific activities. Furthermore, certificates issued by educational institutions, such as community colleges and universities, may be sanctioned by or created in partnership with associations and governments that do not have the means to create training programs themselves.

Now What: Citizen Science, New Collaborations, and Global Networks

Citizen science can be considered a nonprofessional applied science. It is an example of a new type of collaboration between scientists and lay people with an interest in science. Citizen scientists are poised to make an impact on ecological issues around the globe by gathering, documenting, and disseminating massive amounts of data that formally trained scientists can use in their research. The information collected by citizen scientists can be used on many different scales, from an individual park, to a neighborhood, to a district, to a city, to a region, and even nationally and globally. Just as important, scientists and professionals are beginning to make better use of the culture-based knowledge that comes from experience, relationships, and traditional peoples. Citizens impacted by policies and planning decisions are increasingly encouraged to participate in the decision-making process to share their firsthand knowledge of places and user needs. Focus groups, interviews, design charettes, public meetings, and resident surveys to solicit input are all used to gain insight from the people who know a place the best.

New collaborations in education are important to gain different perspectives and problem-solving approaches by connecting various knowledges (Schön 1983). Many colleges and universities are adjusting program offers to include specialty courses, certificates, and programs that are multidisciplinary but focus on an area of expertise. These programs are adjusting to the demand for particular positions that can work in new and ever-changing fields; as cities grow and change, there is a need for new approaches to problems. Climate change, pollution, rapid growth, deteriorating infrastructure, and social changes are challenging cities to come up with innovative solutions that are science- and technology-based, yet respond to social needs. For example, socioenvironmental perspectives are needed for cities that are developing their resilient city plans for environmental stressors such as flooding, sea level rise, and extreme weather events. The issues we are facing in cities today have become more complex than a single profession can address; diverse professionals working in collaboration are required to devise solutions to most of these "wicked problems" (Rittel and Webber 1973).

Challenges being faced today on a global scale include the need for a deeper understanding of cross-scale human system dynamics so that local initiatives can be aligned with global needs. The interactions between people and ecosystems can be unpredictable; as they evolve together, we need adaptive management strategies to deal with them. Ecosystems change, as do the people who attempt to understand and manage the system. As resource managers make decisions within a framework of existing policies and partial knowledge, scientists attempt to understand and communicate system dynamics (Gunderson, Holling, and Light 1995). Working in collaboration allows professionals to better adapt systems and policies and to bring innovation to institutions traditionally set in their ways, thus making science-based decisions rather than taking remedial action or shifting action to the political arena.

Global networks are needed to solve global problems. Global and national programs and governments are taking a multidisciplinary team approach to making decisions and writing policy, particularly around issues that affect us on a planetary scale. The United Nations (UN) has organized countless summits and meetings and has provided frameworks, such as Millennium Development Goals, superseded by Sustainable Development Goals, to guide nations in taking actions that impact the globe as much as their citizens. In the educational realm, the United Nations Educational, Scientific and Cultural Organization (UNESCO) is a leader and has spearheaded initiatives, particularly in the developing world, that also have global impact. More recently, with the advent of climate change, the Intergovernmental Panel on Climate Change (IPCC) has been addressing climate change issues from a scientific perspective, analyzing impacts and providing guidance for mitigation and adaptation strategies that, if adopted by individual nations, can have global impact. Citizen scientists have become active participants in this global enterprise and are changing the way scientists approach research, particularly in remote places (Fritz et al. 2019; UN 2019c).

Citizen Science Project: #FloodMiami

Catch the Tide Miami—Peeking at Future Sea Levels is a project to better understand flooding in the Miami area. The project is led by researchers at Florida International University and was developed to help Miami and other coastal cities that are struggling with flooding related to sea level rise, to better document flooding events. FIU scientists will use the information to provide municipalities with knowledge that can lead to better solutions and actions for flood prevention, particularly during king tides and hurricane season. Participants take photographs of flooded streets, parking lots, and properties and geolocate the information with GPS; they can also add additional information before saving the record. The primary interest is in king tides, because they offer an opportunity to look at areas that will be impacted by future sea levels. The GIS Center and Extreme Events Institute will use the information to validate and improve models for sea level rise research by using the photos to map flood extents and cross-check them with model predictions (SciStarter 2019c).

Project: Living Breakwaters Project

The Living Breakwaters Project, a joint urban and ecological endeavor in Jamaica Bay, New York, by the SCAPE design firm, was a collaborative project to create "oyster-tecture," a living breakwater for oyster habitat that would also reduce risk of damage in powerful storms. Scientists used hydrodynamic modeling and geographic information systems (GIS), plus aerial mapping and sediment collection to test ideas and concepts for an ecological infrastructure. Engineers, marine biologists, hydrodynamic modelers, and graphic designers worked to bring together concepts that connected community, ecology, and infrastructure. The team also worked with regulators, fishermen, and local, state, and federal agencies to create and analyze resiliency goals. The breakwater will be seeded with oysters from the Billion Oyster Project at the Harbor School, where all curricula relate to New York Harbor. Thousands of students across the city have been involved in the project. Community volunteers participated in fuzzy rope weaving to create habitat panels, 14 of which were installed and monitored by scientists from Brooklyn College. In addition to creation of oyster habitat, the shoreline will be replenished as the oysters grow to make recreational beaches and shallow water habitat for wading birds and marine life (Orff 2016, 198–200, 210).

29

Landscape Certification Programs

Certifications that confirm the use of sustainable practices and design for landscapes and buildings have become very popular. When some products began to make doubtful claims of being "green" or sustainable, third-party organizations (those with no association to the company) were established to verify those claims. Third-party organizations can be for-profit or non-profit if they are affiliated with a government or educational program. Two types of certification are common: people with special training can attain professional certificates to validate their credentials, or landscapes and buildings can be certified by meeting established criteria for sustainability. These tests and certifications are often developed by organizations and people who specialize in the science and practice of urban development and urban ecology, such as urban planners, landscape architects, architects, and other environmental scientists. Typical landscape and building certification programs are designed to encourage, regulate, or enforce specific development and maintenance practices by offering a distinguished award or certification for meeting specified sustainable criteria. Science-based criteria used to create the benchmarks for the award often focus on water conservation, soil and tree protection, vegetation use, energy savings, and other construction practices and landscape behaviors that protect the environment. A certification plaque or marker is usually awarded to project designers and/or owners who often use it as a means to secure tax incentives and as a marketing tool to promote the project. Certificates are also offered for materials used in construction and landscape projects. Third-party certification programs are used to test product claims for sustainability or environmental safety. The U.S. Environmental Protection Agency endorses about 30 different third-party certification programs that test products, practices, or buildings that make sustainability claims (EPA 2020).

Most certification programs are either developed, sponsored, or endorsed by private companies and/or government agencies at the national, state, and local level and can be either mandatory or voluntary depending on how they are designated by the agency utilizing the program. For example, local governments may require mandatory participation in a certificate program from a government or nonprofit organization as part of their development process, while others may give developers the option to participate by offering financial incentives. Some programs are created through environmental organizations, such as the National Audubon Society or educational institutions such as land-grant universities that develop and administer programs funded by government agencies. Government agencies that regulate natural resource use, including water management districts, often develop in-house programs that offer incentives, including reduced permit fees, to developers who follow their guidelines and receive certification.

What: Certification Programs and Urban Ecology

Certification programs can be an important component of various strategies, policies, laws, and social marketing campaigns to mandate or encourage good environmental practices in development and maintenance of urban green spaces. National-level programs include the U.S. Green Building Council Leadership in Energy and Environmental Design (LEED) certification, and the Sustainable Sites Initiative (SITES) certification. LEED is the nationally recognized leader in the design and construction of high-performance green buildings (including green roofs and green walls), but the program also has nine other rating systems that cover all aspects of development, including community and neighborhood landscape standards. The SITES certification was developed exclusively for land development and urban landscapes and focuses on the protection and improvement of environmental quality in cities.

The LEED program is a private, nonprofit organization that has developed a set of prerequisites and credits for a building project to achieve certification, at either the silver, gold, or platinum level. Two LEED programs that pertain to urban landscapes include LEED for Cities and LEED for Communities. These programs help communities create sustainable plans for natural systems that encourage resilience and natural resource protection, including water efficiency programs to help cities achieve net-zero water. The certification process includes credit for green spaces, resource conservation, light pollution reduction, and resilience planning (USGBC 2016).

The SITES program was developed by the American Society of Landscape Architects, the Lady Bird Johnson Wildflower Center at the University of Texas at Austin, and the United States Botanic Garden. The science-based program used research from professionals in soil, water, vegetation, and human health to create benchmarks to guide, measure, and recognize sustainable landscape practices (SITES 2009, 8). The rating system includes a variety of benchmarks that include criteria for site selection and assessment, and site design standards for water, soil, vegetation, and construction materials. Other site design strategies are for human health and well-being, including construction, maintenance, and monitoring activities. The goal of SITES is to highlight the importance of ecosystem services in urban areas by transforming land development and management practices that enable natural and built systems to work together for a sustainable site. Benchmarks are specific and measurable criteria for site sustainability that recognize regional differences, including social, cultural and economic differences, and they are targeted toward decision makers and those who design, construct, operate, and maintain landscapes (SITES 2009, 5). SITES can be applied to large-scale open spaces such as parks, botanical gardens, streetscapes, and neighborhoods and to small-scale spaces, including residential yards and commercial and educational campuses (SITES 2014, xiii, xiv).

Gaining certification by meeting LEED and SITES criteria is mandatory for some federal and state projects and the criteria are sometimes used for community projects. Other federal- and state-level programs are often developed through government offices such as environmental protection agencies (EPAs) and are often affiliated with university extension programs or governmental agencies that have jurisdiction over use of natural resources, such as water management districts. Use of these programs may be backed by state statutes or laws; some are mandatory, but most are voluntary and sometimes included in mandated practices for community development. Social marketing messages are the primary means of promoting these certification programs, often through county extension programs.

So What: How Certification Programs Benefit Urban Ecology

Regardless of whether a program is mandatory or voluntary, they all have the common goal of protecting the environment through the application of select science-based activities backed by research or empirical evidence. Certification programs are particularly applicable to environmental issues in urban areas because they set criteria for development and maintenance activities that benefit the ecology of sustainable green spaces.

Protection of natural resources is included in many of the SITES criteria that specify site selection and preparation for construction. For example, soil protection strategies include appropriate site selection, such as limiting development on sensitive land and prime farmland. Criteria also require creating a soil management plan and protection of natural resources, especially in streams, floodplains, and wetlands. Other water-related criteria include reducing the use of potable water for irrigation, managing stormwater on site, designing stormwater features as amenities, and providing clean water for recreation activities. Vegetation criteria include preserving existing plant biomass on site to improve greenhouse gas absorption and minimize heating and cooling costs, using appropriate, non-invasive, native plants that use fewer resources, and managing invasives to prevent competition with native plants (SITES 2014, 2–51).

Construction criteria focus on using sustainable materials and maintenance practices that protect soil, water, and air quality. Recommended activities include the use of salvaged or recycled materials to reduce material consumption, using certified wood and regional materials to reduce transportation pollution, and using low volatile organic compound (VOC) materials to moderate VOC pollution in the ground-level ozone layer. Maintenance activities that protect the environment after development are included in the criteria to maintain the sustainable qualities of the site, including recycling organic matter (composting) to improve soil quality and minimizing pesticide and fertilizer use to help protect water bodies and aquatic ecosystems. Other maintenance criteria include using energy efficient outdoor equipment and minimizing the use of equipment that generates greenhouse gas and pollutants (SITES 2014, 58–135).

SITES also includes criteria related to social issues, including protection and maintenance of cultural and historic places that will help increase tourism employment and enhance attachment to land for a strong sense of stewardship. Creating community cohesion and providing opportunities for outdoor physical activity are also social goals that contribute to the

FIGURE 29.1. Perth, Australia: the use of drip irrigation is one of several criteria for most landscape certification programs.

health of the community. A new policy for many certification programs is to develop a plan or strategy to monitor and report performance over time as the project ages to improve the body of knowledge for long-term sustainability practices. Criteria include dissemination of results in professional magazines and peer-reviewed journals or at professional conferences to help strengthen certification programs (SITES 2014, 70–117).

SITES and LEED are national programs that can be adapted for regional context, but some states use their own certification programs to focus on regional or local environmental issues in the state. Several states have programs through cooperative extension, such as the Florida-Friendly Landscaping (FFL) program of the University of Florida, the Carolina Yards Program through Clemson University, and the Monterey Bay Friendly Landscape Certificate, funded in part by the California State Water Resources Control Board. These programs typically certify homeowner yards and commercial landscapes with checklist criteria based on principles promoted by the program, such as watering efficiently, using mulch, reducing fertilizer use, and protecting waterfronts from stormwater runoff. Criteria are set for elements in the landscape, such as using a drip system (fig. 29.1) and a rain shut-off device for irrigation and a specified mulch depth for plant beds. Homeowners who participate in programs request

FIGURE 29.2. Gainesville, Florida: yard signs are given to homeowners who participate in the Florida-Friendly yard program and meet certification requirements.

an inspection of their yard and are given a yard sign to show their yard is certified by the program (fig. 29.2) (FFL 2017). Many homeowners who participate in certification programs such as the Florida-Friendly Landscaping program feel it is their social responsibility to use environmentally friendly landscapes, but they also engage in certification programs to gain recognition, show community pride, and influence their neighbors.

Although little formal research has been done on the best types of certification/reward programs, evidence from some programs show they are successful, and that recognition is a powerful tool to reduce environmental impacts. One example is Harvard University's Green Carpet Awards, an annual recognition event that honors students, staff, and faculty who support the university's sustainability efforts. The event also awards behavior change by participants who adopt sustainable practices such as recycling and reduced food waste (Baier 2011).

Now What: Problems and Solutions with Certification Programs

Although the LEED program has received many awards, it has also been criticized for not actually creating energy efficient buildings. One analysis found that builders often target the lowest or easiest criteria to get the minimum certification, which often does not translate to better energy efficiency. Data from several New York City buildings that were LEED certified showed they actually performed worse than other noncertified buildings. The reason may be that LEED has so many credits, including for things such as bicycle racks, that the importance of the energy efficient credits is diluted (Rosiak 2013). One problem is the lack of independent or meaningful research into LEED's effectiveness, but more than 200 states, cities, and federal agencies now require LEED certification for public buildings, and nearly 200 give LEED builders tax breaks and other incentives. Despite this, many environmentalists give LEED credit for expanding the use of green practices and transforming the building industry. Another problem is that LEED certification was awarded before occupancy, based on energy use projections, not on actual energy or water use. To rectify this, the new version of LEED, which became mandatory in 2015, requires building operators to develop a plan for running a building efficiently and reporting energy and water use for five years (Schnaars and Morgan 2012). The LEED for Cities and Communities program requires communities to share performance data to measure and track progress toward their goals. Most programs, including SITES, now have criteria for third-party monitoring for the performance of the sustainable design practices included in the certification checklist. The goal is to evaluate performance over time to improve knowledge and improve the efficacy of the criteria used in certification programs. Participants receive credit when they complete and report the evaluation, and negative reports do not affect other credits earned (SITES 2014, 114).

Some homeowners have experienced conflicts between using programs to certify their yards and meeting HOA landscape code requirements. Recent cases have emerged in which HOAs in Florida have gone to court with homeowners for installing a Florida-Friendly landscape with bahiagrass that did not meet the HOA turf requirements for St. Augustine or zoysia turf (Kassab 2016). To help homeowners and HOAs resolve conflicts and to encourage the use of FFL principles, the Florida legislature defined Florida-Friendly Landscaping in Florida Statutes section 373.185 and adopted Senate Bill 2080 (2009) that allows homeowners to install FFL yards, but it does not excuse homeowners from going through the normal review approval process with the HOA board for landscape modifications (Momol et al. 2014). As a result, several legal challenges are currently going through Florida courts by homeowners denied approval to use FFL landscapes (Manning-Hudson 2012).

Citizen Science Project: Monitoring and Evaluation

Although citizen science (CS) projects have not been used specifically to investigate landscape certification projects, a recent study investigated the use of citizen participation in monitoring and evaluation of the landscape design process. Lessons learned from this study could be applied to the monitoring and evaluation of certificate programs for urban landscapes. Currently post-occupancy evaluation and landscape performance research are used to evaluate projects, but they are not used extensively in practice because of time constraints, lack of resources, and lack of professional interest in the process. Key considerations for including citizen science activities for monitoring and evaluation during the design process include: 1) CS monitoring and evaluation can contribute to different stages of design and be incorporated iteratively into the process; 2) monitoring strategies should/could include several approaches from disciplines related to landscape architecture and urban ecology; 3) CS-based monitoring can ensure public participation throughout the process; and 4) CS can increase the capacity for monitoring and evaluation of landscape projects through increased data collection, knowledge-sharing, and participation in design and planning. Citizen participation can make design and designers more responsive to local needs, encourage stewardship in participants, and directly benefit decision-making for design projects (Livingstone 2017).

Case Study: Phipps' Center for Sustainable Landscapes— Certified SITES Project

The Phipps Center, located in Pittsburgh, Pennsylvania, is a 2.9-acre urban garden that was formerly a brownfield where the City of Pittsburgh Public Works Yard was located. Before construction of the center, the entire site was paved over and there was no vegetated land cover. The site underwent a remarkable transformation before the center was opened in 2012, receiving four of the highest green building standards, including Platinum Sustainable SITES, LEED Platinum, WELL Building Platinum, and

Living Building Challenge. The facility is net-zero energy and net-zero water, producing all of its energy and managing all stormwater on-site with rain gardens, bioswales, a lagoon, and pervious asphalt. The facility also treats all sanitary water with a closed loop system that uses a settling tank, constructed wetland, underground sand filters, and UV filters. The site uses the sun, earth, and wind to light, heat, and cool the interior. Features include 1.5 acres of green space with more than 100 native plant species in open meadows, oak woodlands, and wetland plantings. The ethnobotanical-inspired permaculture green roof uses plants with documented historic food, medicine, and craft uses. A 4,000 ft^2 lagoon is fed by roof runoff from the conservatory and is home to native fish and turtles. A 60,000-gallon underground rain tank annually stores about 500,000 gallons of rooftop runoff for irrigation. Visitors to the center can learn about native plant communities and the role of green infrastructure in improving water quality and observe native wildlife in the regenerative landscape (Piacentini 2019).

30

Urban Ecology of Future Cities

Cities first appeared in Iraq about 6,000 years ago, and since that time cities have become the dominant form of human settlements. It is estimated that 3 out of 5 people on the planet will live in cities by 2050, with the most growth in the least urbanized continents of Asia and Africa. Today, megacities such as Tokyo, Shanghai, and Delhi support about 20 to 30 million residents, and currently 28 megacities exist around the globe, each with a population of 10 million or more people (Bliese et al. n.d.). Many of these and other cities are grappling with the future, struggling with current problems and looking to avert future problems. New cities for twenty-first-century people are only beginning to be built in many parts of the developing world, offering great possibilities and great challenges. Population growth and climate change will require rethinking many of our political, economic, cultural, and social dynamics in cities, especially in the context of ecological and environmental risks to nature and people. Pollution, social and environmental justice (fig. 30.1), health, energy, transportation, biodiversity, water supplies, and impacts of climate change such as floods, severe weather, droughts, and sea level rise are all concerns for urban areas. Urban ecologists, planners, architects, landscape architects, and civil engineers are embracing new paradigms and concepts for resilient cities, including ecological urbanism, biophilic design, political ecology, and biosphere reserves to create a vision for future cities (Heise 2019).

When asked about their vision of future cities, most people imagine cities of science fiction: domed colonies on Mars, structures of steel and light floating in space, glass bubbles under the sea, and networks of underground cities with light and air tubes. Cities such as these exist only in the imagination and in fictional stories of science and fantasy. Ventures in space and in underwater and underground habitats have been tried with varying success. The Mir Space Station, where a cosmonaut has stayed a record 437 days, and Tektite I, an underwater research station

FIGURE 30.1. Harbin, China: abandoned fishing boats on the Songhua River are emblematic of pollution and environmental justice problems that need to be solved in future cities.

where aquanauts lived for 58 days, are examples of artificial habitats that can support a few people for a short time, with breathable air and waste management being primary limiting factors. Perhaps the most successful alternative living habitats have been underground structures, where natural light and air can be accessed. Many cities already have underground facilities such as shopping malls, data centers, and subway systems; however, these systems are not large and self-sustainable; they rely on air, water, and light from above, which makes ecosystem services just as important below ground as above. Helsinki, Finland, for example, has several existing underground facilities, and the city has plans to greatly expand its underground with a strategic Underground City Plan, including construction of 200 underground structures, including apartments and public spaces (Mancebo 2017).

Future cities for the twenty-first century are being built today, with today's technology and planning practices, so a more useful question to ask about future cities is what qualities humans need and *want* in future cities. Most people would likely (hopefully?) say we need sustainable cities that balance consumption now and in future generations to be a truly long-lived city, others might say we need strong (resilient) cities so we can survive storms and other environmental stresses and carry on with our daily lives, and most people would say they *want* cities they can live in. Livable cities support all inhabitants equally and provide ecosystem services and functionality so all citizens can be prosperous and productive. The challenge for urban planners and decision makers is to connect the concepts of sustainability, resilience, and livability to city planning, and the challenge for scientists is to define and specify the concepts so they are operational, meaning they can be translated to policy and strategies with well-defined actions that can be used for development plans. In this way both the thinkers and the doers are connected through useful, usable knowledge (Maddox 2013).

Ecological Cities

Urban planners and city designers believe that nature and ecology will be the answer to many of the population growth and climate change pressures of tomorrow. Planning for the future includes many different models: resilient cities, sustainable cities, national park cities, sponge cities, reconciliation cities, urban biosphere reserves, foodtopias, and zoöpolis (cities designed for humans and animals). All of these models are based in ecology and ecosystem services and address various natural and built features that, combined, make a city sustainable and resilient. The April 2019 special issue of *National Geographic* magazine features articles on cities, including a vision for the future created with the help of the architectural and urban planning firm Skidmore, Owens & Merrill (SOM). SOM described the components of a future city that are based on ecology, smart buildings, and city form, including connected urban hubs separated by green spaces and waterways, smart buildings that are energy efficient, and self-contained neighborhoods with mixed income communities. Other concepts include future cities where waste is a resource, food is grown locally, and new forms of transportation increase mobility. Together these concepts create future cities that are sustainable, livable, energy efficient, and socially diverse (National Geographic 2019).

Resilient Regions for City Planning

Planning for large cities in the future will take a different approach to spatial form and growth options. Future cities are no longer envisioned as a central core that spreads organically with population growth; large cities will fragment into small dense urban hubs dispersed throughout a region and connected by high-speed rail systems. *Urban Region Plans* are a tool that could be used to determine the most appropriate and/or inappropriate areas for new hubs, for agriculture areas and industrial centers, and for water-supply protection areas (Forman and Wu 2016, 610). Hubs are located to work with the regional ecology, including preserving landforms such as hills and valleys, protecting undisturbed natural areas such as wilderness and forests, and protecting rivers, streams, and lakes. This approach to biomorphic urbanism—having the form of a living organism—is based on several principles: resiliency zones, outdoor recreation areas, protected natural wilderness, city wilderness areas, sustainable agricultural areas, coastal protection zones, and transit corridors to connect distant hubs. Resiliency zones mean no building in flood-prone areas and using natural and human-made structures to manage stormwater. Protected wilderness areas surrounding the region are closed to development and other uses, while the city wilderness areas serve as wildlife habitat and are open to nature-based recreation activities. These areas also serve as flood management facilities built to handle occasional inundation but are normally accessible for recreation. Agricultural areas are set aside on the perimeter of the hubs to limit transport distance for food supplies, and development is set back from coastal areas to protect against sea level rise. High-speed transportation systems connect the hub centers to reduce sprawl and make employment centers easily accessible (National Geographic 2019).

Natural Areas for Future Cities

Urban hubs are small to mid-sized cities designed with nature in mind. Components of an ecology-based urban hub are targeted to protect resources and limit impacts on the ecosystem. For example, natural water courses that run by or through the hub, such as rivers and streams, are protected with green riparian habitats along the shoreline rather than building to the edge with bulkheads and seawalls. Water management in the city includes flood protection strategies with natural vegetated barriers that create new marine habitats. Wetland and natural waterway restoration is also used for flood protection and wildlife habitat. Facilities for collecting, cleaning, and recycling water drive the form for much of the city infrastructure. Protecting upland water sources within the resilient regions, capturing and cleaning stormwater, and restoring wetlands are strategies

FIGURE 30.2. Harbin, China: sponge city wetland parks increase storage and percolation capacity in a new high-density residential development.

for water management that include using sponge city practices to increase every opportunity for water percolation and storage (fig. 30.2). All parks and green infrastructure are designed to percolate water in bioswales to recharge the aquifer. Stormwater structures include next-generation green roofs that help collect water and provide opportunities for energy reduction and small-scale farming. Many cities today are already using new and innovative green systems for stormwater management. Chulalongkorn University Centenary Park in Bangkok is an example of a park designed to reduce flooding and mitigate disaster risk. Bangkok is sinking at the rate of 0.8 inches per year among rising sea levels, and it is estimated that the city could be completely submerged by 2030. Centenary Park extends for 11 acres along a sloped 1-mile stretch of road. Water movement begins with a green roof at the high point then flows through wetlands and filtration ponds to a retention pond at the lowest point. Three underground tanks store rainwater from the green roof for irrigation, and visitors can ride stationary water bikes located along a pond to aerate the water. The park is an example of green infrastructure that can help mitigate climate change stressors while providing an active green space in a dense urban area (Mortice et al. 2019b, 170–171).

While natural regional vegetation is preserved around the hubs, the emphasis is on "useful" vegetation in the hub core. Street trees are selected for maximum shade to reduce the urban heat island, and edible plants appear on rooftops, in backyards, in school gardens, and in urban farms and gardens. High-yield vertical gardening and hydroponics allow maximum yield for small spaces, and selected plants provide habitat for pollinators and birds. Native and drought tolerant plants are used for landscaping to reduce irrigation needs, and vegetation is also used to cover buildings with green roofs and green walls to reduce energy needs. A central *Environmental Center* will be used to monitor ecosystems using standard indicators for air, water, and soil health. Clean energy and clean air systems such as bladeless air turbines on tall buildings and green ventilation systems will help protect air quality (National Geographic 2019).

Structures in Future Cities

Smart buildings will incorporate natural elements with modular spaces for housing, industry, and business needs. The main concept of smart buildings is renewable energy, where buildings generate as much energy as they consume and share energy with others for a self-sufficient city. Smart buildings include features such as sky gardens, natural lighting from bioluminescent materials, bladeless wind turbines, and solar walls and windows to generate energy. Sky gardens are built into different levels throughout tall structures to provide social areas and help promote airflow in the buildings. Bosco Verticale, the vertical Forest Towers in Milan, Italy, are an example of sky gardens in use today. The residential complex has two towers, 160 and 250 feet tall, and is a pilot project for sustainable buildings. Balconies on every level hold more than 100 different plant species, including 800 trees 10–30 feet high, 5,000 shrubs, and 15,000 ornamental plants. Plants are in concrete containers irrigated by a central system, and maintenance workers hang from cranes on the top of the buildings to maintain plants from the outside. Vegetation of each tower is equal to the number of trees found in five acres of forest and provides microclimate, humidity, and oxygen and absorbs CO_2 and pollution. Plants also support insects and birds; around 20 different bird species have built nests in the towers (Bianchini 2019). Buildings are also designed and constructed to enhance social life and well-being for everyone. Modular units provide flexibility in uses and less construction waste, and interiors are fully accessible for all mobility needs and designed for intergenerational housing so all ages can enjoy the benefits of social housing. Grouped services such as on-demand delivery using smart refrigerators and kitchens will make multiple unit deliveries in one trip for more efficient transportation of goods and to reduce surface traffic. Recreation, arts, and entertainment are also shared with virtual and augmented reality (National Geographic 2019).

Food and Waste Management

In future cities, food is produced using sustainable practices such as organic farming and hydroponics. Underground farming beneath homes and offices and vertical farming on different floors of tall buildings will produce food for small local markets distributed around the city. Underground farming is one concept for a future city that shows promise. Similar to underground, indoor farming is already being used with success in Japan. Spread Company has a new facility, Techno Farm, that can grow 30,000 heads of lettuce a day, a yield of 648 heads per 10 square feet annually, compared with 5 heads at an outdoor farm. Cultivation racks with light-emitting diodes and temperature, humidity, and irrigation controls use 1% of the volume of water needed outdoors per head of lettuce, significantly reducing waste. Sealed rooms also protect vegetables from pests and diseases, eliminating the need for pesticides. Currently 60% of the facilities in Japan are unprofitable or government subsidized because of energy costs to operate the farms; however, Spread Company has spent years refining their systems with custom designed lights that use 30% less energy, and the company is now profitable. Companies in Japan are planning to export their farming system to more than 100 cities worldwide in places such as China, where climate and pollution concerns create a demand for year-round "clean" food (Takada 2018).

Future cities will also manage waste as a resource, including for energy production and buildable landfill. Wastewater will also be treated for irrigation and consumption (National Geographic 2019). Data from the European Union Data Centre on Waste found that in 2010 around 60% of waste generated in the EU consisted of mineral and soil waste from mining and construction and demolition activities. Minerals and soil could be used for buildable landfill, as opposed to waste landfills, which are not appropriate for building because of the release of methane gas from decomposing waste. Other waste products, including metal, paper and cardboard, chemical and medical waste, and animal and plant waste range from 2% to 4% of the total waste generated. Electronics is the fastest-growing waste stream in the EU, which has set ambitious targets for waste management with the *Waste Framework Directive*, which outlines the management hierarchy, starting with prevention, then reuse, recycling, recovery, and disposal. The goal is to use waste as a resource and minimize landfill and pollution. Environmental issues include air pollution, greenhouse gas emissions, and soil and water contamination that affect many ecosystems and species, including crops grown in contaminated soil and fish that ingest toxic chemicals. Incinerating or recycling waste can be used to produce heat or electricity to replace energy from coal and other fuels. Turning waste into a resource is a key objective of the EU's *Roadmap to a Resource Efficient Europe*. A study from 2011 found that improved management between 1995 and 2008 significantly reduced greenhouse gas emissions as a result of lower methane emissions from landfills, and by 2020 if all EU countries meet their waste diversion targets an additional 68 million tons of CO_2 emissions would be cut from the waste life cycle (European Environment Agency 2016).

Transportation

Urban hubs are designed to be compact and connected to other hubs within a region with high-speed, energy-efficient transportation. Social transit will include high-speed rail stations that are centers for community activities. Mixed-use districts make neighborhoods walkable and efficient, with amenities and services within a 10-minute walk. The smart street concept means streets are mostly car free and designed primarily for pedestrians and public transportation, which will also include remotely programmed drones to fly humans above the city and skyways to connect buildings and reduce street-level traffic. Transportation centers around the city will be central to daily life, with meeting and eating places, markets, galleries, and cultural events (National Geographic 2019). Efficient transportation will increase mobility through automated technology with high-speed rails (HSR), light rails, commuter trains, and pedestrian and bicycle networks all connected to central transit hubs. High-speed rails are not new, but they are not widely used today; however, with the new urban form of hub cities, they will provide more efficient transportation between hubs. Several countries already have networks for high-speed rails with maximum speeds of 185–220 mph; the top five include cities in China, Spain, Japan, France, and Germany. Japan was the first to have a high-speed rail system; the Shinkansen or "bullet train" was built in 1964, but China recently surpassed the rest of the world in miles of rail lines and by 2025 expects to have more lines than the rest of the world combined. Environmental benefits include energy savings; by reducing the number of vehicles on the road, HSR is four times more energy efficient than driving cars and nine times more efficient then flying. HSR also lowers greenhouse gas emissions. For example, the California High-Speed Rail Authority estimates that by 2040 the new California system will reduce auto traffic by 10 million miles a day and over a 58-year period reduce miles by more than 400 billion miles of travel. The rail authority also estimates that starting in 2030, daily flights can be reduced by 100–200 a day. HSR is seen as a triple bottom line: economic, social, and environmental, for sustainability (Nunno 2018).

Future Cities Today

Many of the ideas and technology for future city concepts are available and used primarily in developed countries, but no individual city has incorporated all or even most of the concepts, since they would require considerable redevelopment. However, more developing countries are rapidly building new cities and will be the areas of largest urban growth. China is one example of a developing country with ambitious national plans to rapidly develop more urban areas and shift people from rural areas to cities to keep their economy growing. In a 34-year span (1978–2012) China's population in cities increased from 18% to 53%, and although they have employed a few of the strategies for future cities, they have also missed the mark on some important issues where unintended consequences and high ecological stress can serve as examples for other countries. To remedy the problems, China has adopted a new program, the *National New-Type Urbanization Plan*, that emphasizes a people-centered approach for better welfare and well-being, with the goal of fixing problems experienced in the last 30 years of rapid urbanization. For example, to solve problems in transitioning rural people to urban areas, the government aims to increase the income of new urban residents to match the income of existing residents; however, this may not be without consequences as the anticipated new levels of consumption will increase CO_2 emissions that cannot be offset by the planned low-carbon strategies. Food scarcity is another concern; about half the urban growth has been on arable land, and although the government introduced strict regulations mandating that cities secure the same

FIGURE 30.3. Harbin, China: using every available bit of soil to grow edibles, a small patch of leeks grows in a roadway median.

amount of agricultural land elsewhere, cities have turned to other nonagricultural land to develop, destroying wetlands and lakes and carving up mountainsides for development. The city of Wuhan, once famous for its lakes, has destroyed more than 70% of them to reclaim land for development. In addition, rural residential lands have been expropriated to force urbanization on rural residents; while this strategy has helped cities grow at low cost, a lack of security has made rural residents reluctant to give up their rural land quota, preferring to leave it uncultivated, increasing the food security dilemma (fig. 30.3). Residential land use has also increased rapidly in rural areas, further reducing agricultural lands. Rapid urban growth has also left about two-thirds of China's cities with water shortages due to severe surface water and groundwater pollution, and air and water pollution have also been blamed for respiratory and cancer deaths in some areas that are 20–30 times higher than the national average. Some cities, however, are doing well by promoting public participation in the planning process and aligning their development goals with social, economic, and environmental goals; seeing urbanization as an opportunity for sustainable technology, for example, by improving building codes for more green buildings. In the past a major problem has been the lack of flexibility in national plans to accommodate local practices; however, the focus on sustainability and well-being in the new plan should help alleviate some past problems (Bai, Shi, and Liu 2014, 158–160). Although controlling rapidly growing cities with a single national plan has proved problematic, other cities have been successful with local city planning that takes into account local issues and residents' needs. Cities in several countries are currently adopting several of the future city strategies in their buildings and green spaces and have been able to show various benefits. For example, Singapore increased its population by 68% from 1986 to 2007, and at the same time increased green space from 35.7% to 46.5%. One-fifth of the floor area in the city is green certified buildings, and their goal is to have 80% green buildings by 2030 with a 35% reduction in energy use (Lieber 2018).

Future Research for Cities

A new frontier for future cities is research and policy development. The interrelated complexity of urban studies will require research in many disciplines and organizations. Urban challenges include social and environmental issues, such as ecological restoration and protection, environmental justice, economic opportunity, green infrastructure development, resiliency and sustainability, and ecological planning. Important areas needing investigation include material usage, the functionality of urban ecosystems, the role of political systems and social norms, developing migration settlements, public health concerns such as disease vectors, and innovation in urban planning. Unfortunately, no central organization or agency exists to support urban researchers, which makes it difficult to organize and present a cohesive set of strategies that are useful to policy makers. Scientists come from many disciplines and are spread among academic institutions, government agencies, nongovernmental organizations, and community-based organizations, which makes it difficult to coordinate research strategies. Steps have been suggested by leading urban ecology scientists on ways to improve the science of cities, including forming a global urban scientific body, spreading knowledge globally, boosting funding, supporting transdisciplinary research, and improving access and linkages in science-policy areas. A global urban science body is needed for science-policy interaction and cross-city learning, an area where citizen scientists can make a significant impact. Creation of a science body should include input from scientists and people who have urban knowledge, such as civil servants and citizens. National governments, private foundations, and development banks can and should help fund research grants and engagement activities with communities and policy makers who need better access to studies. This global urban science body can be facilitated by urban scholars who play a role in policy development, particularly through synthesis of transdisciplinary research. The goal is to scale up urban research, involve all stakeholders, including citizens, and develop a system to direct and implement urban planning policy for better cities based on ecology (McPhearson, Parnell et al. 2016a).

Bibliography

Adams, Clark E., and Kieran J. Lindsey. 2011. "Anthropogenic Ecosystems: The Influence of People on Urban Wildlife Populations." In *Urban Ecology, Patterns, Processes, and Applications*, edited by Jari Niemelä, Jurgen Breuste, Thomas Elmqvist, Glenn Guntenspergen, Philip James, and Nancy E. McIntyre, 116–128. New York: Oxford University Press.

Adams, Jennifer, David Greenwood, Mitchell Thomashow, and Alex Russ. 2016. "Sense of Place." Accessed October 27, 2019. https://www.thenatureofcities.com/2016/05/26/sense-of-place/

AgEdLibrary. n.d. "Determining the Importance of the Horticulture Industry." Accessed October 18, 2019. http://tuscolaagriculture.weebly.com/uploads/8/3/8/9/8389114/importance_of_hort.pdf

AIA (The American Institute of Architects). 2019. Accessed November 22, 2019. https://www.aiatopten.org/node/140

Alberti, Marina, and John M. Marzluff. 2004. "Ecological Resilience in Urban Ecosystems: Linking Urban Patterns to Human and Ecological Functions." *Urban Ecosystems* 7: 241–265.

Alexander, Ernest R. 2005. "What Do Planners Need to Know? Identifying Needed Competencies, Methods, and Skills." *Journal of Architectural and Planning Research* 22(2): 91–106.

American Bird Conservancy. n.d. "Feral Cats: Consequences for Humans and Wildlife." Accessed July 24, 2019. https://abcbirds.org/wp-content/uploads/2015/05/Feral-Cats-Consequences-for-Humans-and-Wildlife.pdf

Anderies, John M. 2014. "Embedding Built Environments in Social-Ecological Systems: Resilience-Based Design Principles." *Building Research & Information* 42(2): 130–142.

Andersson-Sköld, Yvonne, Jenny Klingberg, Bengt Gunnarsson, Kevin Cullinane, Ingela Gustafsson, Marcus Hedblom, Igor Knez, Fredrik Lindberg, Åsa Ode Sang, Håkan Pleijel, et al. 2018. "A Framework for Assessing Urban Greenery's Effects and Valuing its Ecosystem Services." *Journal of Environmental Management* 205: 274–285.

Archdaily. 2012. "Via Verde / Grimshaw + Dattner Architects." Accessed December 16, 2019. https://www.archdaily.com/468660/via-verde-dattner-architects-grimshaw-architects

Arizona Water Awareness. n.d. "Arizona Water Awareness Month, Rebates for Water Conservation." Arizona Department of Water Resources. Accessed September 23, 2019. http://waterawarenessmonth.com/rebates.php

Aronson, Myla, Madhusudan Katti, Paige Warren, and Charles Nilon. 2019. "Comparative Ecology of Cities: What Makes an Urban Biota 'Urban?' Accessed August 5, 2019. https://www.nceas.ucsb.edu/workinggroups/comparative-ecology-cities-what-makes-urban-biota-urban

Atkins. 2012. "Future Proofing Cities, Creating Cities Fit for the Future." Accessed August 29, 2019. https://www.atkinsglobal.com/~/media/Files/A/Atkins-North-America/Attachments/sectors/urban-development/library-docs/brochures/Atkins-future-proofing-cities.pdf

Audubon International ACSP. 2019. "About the ACSP for Golf Program." Accessed November 4, 2019. https://auduboninternational.org/wp-content/uploads/2019/03/ACSP-Golf-Fact-Sheet-2018.pdf

Bai, Xuemei, Peijun Shi, and Yansui Liu. 2014. "Realizing China's Urban Dream." *Nature* 509: 158–160.

Baier, Paul. 2011. "The Power of Rewards and Recognition in Sustainability Programs." Accessed October 4, 2019. https://www.greenbiz.com/article/power-rewards-and-recognition-sustainability-programs

Baker, Fraser, Claire Smith, and Gina Cavan. 2018. "A Combined Approach to Classifying Land Surface Cover of Urban Domestic Gardens Using Citizen Science Data and High-Resolution Image Analysis." *Remote Sensing* 10(4): 537–560. https://www.mdpi.com/2072-4292/10/4/537 . Accessed August 27, 2021.

Balling, John D., and John H. Falk. 1982. "Development of Visual Preference for Natural Environments." *Environment and Behavior*, 14(1): 5–28.

Baltimore Ecosystem Study. n.d. "History of the Baltimore Ecosystem Study." Accessed December 15, 2019. https://baltimoreecosystemstudy.org

Barbosa, A. E., J. N. Fernandes, and L. M. David. 2012. "Key Issues for Sustainable Urban Stormwater Management." *Water Research* 46: 6787–6798.

Barton, Sean. 2017. "Smartphone App Could Reveal How Urban Spaces Affect Our Health and Wellbeing." University of Sheffield. Accessed October 28, 2019. https://phys.org/news/2017-07-smartphone-app-reveal-urban-spaces.html

Beatley, Timothy. 2011. *Biophilic Cities: Integrating Nature into Urban Design and Planning*. Washington, DC: Island Press.

———. 2016. *Handbook of Biophilic City Planning and Design*. Washington, DC: Island Press.

Beck, Travis B., Joe E. Heimlich, and Martin F. Quigley. 2002. "Gardeners' Perceptions of the Aesthetics, Manageability, and Sustainability of Residential Landscapes." *Applied Environmental Education and Communication* 1: 163–172.

Bedimo-Rung, Ariane, Andrew J. Mowen, and Deborah A. Cohen. 2005. "The Significance of Parks to Physical Activity and Public Health: A Conceptual Model." *American Journal of Preventive Medicine* 28: 159–168.

Belaire, Amy J., Lynne M. Westphal, and Emily S. Minor. 2016. "Different Social Drivers, Including Perceptions of Urban Wildlife, Explain the Ecological Resources in Residential Landscapes." *Landscape Ecology* 31:401–413.

Bernard, Michael M. 1965. "The Development of a Body of City Planning Law." *American Bar Association Journal* 51(7): 632–636.

Bianchini, Riccardo. 2019. "The Vertical Forest Towers in Milan by Boeri. Phenomenon or Archetype?" Accessed December 3, 2019. https://www.inexhibit.com/case-studies/the-vertical-forest-towers-in-milan-by-boeri-phenomenon-or-archetype/

The Biodiversity Institute. n.d. "Citizen Science: The Biodiversity Institute." Accessed August 9, 2019. https://www.wyomingbiodiversity.org/index.php/Initiatives-Programs/CitSci

Biron, Carey L. 2018. "Developers Tee Off amid Mass Closure of U.S. Golf Courses." Accessed November 5, 2019. https://www.reuters.com/article/us-usa-land-sport/developers-tee-off-amid-mass-closure-of-u-s-golf-courses-idUSKCN1N64TK

Blaine, Thomas W., Susan Clayton, Paul Robbins, and Parwinder S. Grewal. 2012. "Homeowner Attitudes and Practices Towards Residential Landscape Management in Ohio, USA." *Environmental Management* 50: 257–271.

Bliese, Carol, Lindsey Bailey, Lauren Boucher, Lauren Carlson, Drew Grover, Isabelle Rios, and Pam Wasserman. n.d. "World Population History, Urbanization and the Megacity." https://worldpopulationhistory.org/urbanization-and-the-megacity/

Bonney, Rick, Caren Cooper, and Heidi Ballard. 2019. "The Theory and Practice of Citizen Science: Launching a New Journal." Accessed November 19, 2019. https://theory-andpractice.citizenscienceassociation.org/articles/10.5334/cstp.65/print/

Boone, Christopher G., and Michail Fragkias, eds. 2013. *Urbanization and Sustainability: Linking Urban Ecology, Environmental Justice and Global Environmental Change.* Vol. 3, Dordrecht & New York: Springer.

Boston. 2019. Boston Planning and Development Agency. Accessed December 23, 2019. http://www.bostonplans.org/planning/planning-initiatives

Boundless Biology–Lumen Learning. n.d. "The Scope of Ecology." https://courses.lumenlearning.com/boundless-biology/chapter/the-scope-of-ecology/

Braga, Benedito P. F., M.F.A. Porto, and R. T. Silva. 2006. "Water Management in Metropolitan São Paulo." *International Journal of Water Resources Development* 22(2): 337–352.

Bratman, Gregory N., Christopher B. Anderson, Marc G. Berman, Bobby Cochran, Sjerp de Vries, Jon Flanders, Carl Folke, et al. 2019. "Nature and Mental Health: An Ecosystem Service Perspective." *Science Advances* 5(7): 1–14.

Brennan, Andrew. 2015. "Environmental Ethics." *Stanford Encyclopedia of Philosophy.* Accessed November 9, 2019. https://plato.stanford.edu/entries/ethics-environmental/

Browning, William D., Catherine O. Ryan, and Joseph O. Clancy. 2014. *14 Patterns of Biophilic Design.* New York: Terrapin Bright Green.

Bruner Foundation Inc. 2013. "Inspiring Change: The 2013 Rudy Bruner Award for Urban Excellence." Accessed December 16, 2019. https://www.rudybruneraward.org/wp-content/uploads/2016/08/07-Via-Verde.pdf

Bruyere, Brett, and Silas Rappe. 2007. "Identifying the Motivations of Environmental Volunteers." *Journal of Environmental Planning and Management* 50(4): 503–516.

Bryan, William. 2017. "I See Change." Accessed September 1, 2019. https://www.nasa.gov/solve/feature/i-see-change

Burke, Fintan. 2019. "Emotional Response to City Design Could Guide Urban Planning." *Horizon Magazine.* Accessed October 28, 2019. https://horizon-magazine.eu/article/emotional-response-city-design-could-guide-urban-planning.html

Burns, Matthew J., Tim D. Fletcher, Christopher J. Walsh, Anthony R. Ladson, and Be-linda E. Hatt. 2012. "Hydrologic Shortcomings of Conventional Stormwater Management and Opportunities for Reform." *Landscape and Urban Planning* 105: 230–240.

Byrne, Jason. 2011. "The Human Relationship with Nature." In *The Routledge Handbook of Urban Ecology*, edited by Ian Douglas, David Goode, Kike Houck, and Rusong Wang, 63–73. New York: Routledge.

CACWNY. 2019. Clean Air Coalition of Western New York. Accessed December 14, 2019. https://www.cacwny.org

Cal-IPC. 2012a. "Preventing the Spread of Invasive Plants: Best Management Practices for Transportation and Utility Corridors." Cal-IPC Publication 2012–01. California Invasive Plant Council, Berkeley. Accessed November 6, 2019. https://www.cal-ipc.org/docs/bmps/dd9jwo1ml8vttq9527zjhek99qr/BMPsTransportUtilityCorridors.pdf

———. 2012b. "Preventing the Spread of Invasive Plants: Best Management Practices for Land Managers (3rd ed.)." Cal-IPC Publication 2012–03. California Invasive Plant Council, Berkeley. Accessed November 6, 2019. https://www.cal-ipc.org/docs/bmps/dd9jwo1ml8vttq9527zjhek99qr/BMPLandManager.pdf

Callanan, Pippa L. 2010. "Intrinsic Value for the Environmental Pragmatist." *Res Cogitans* 1(1): 132–142.

CEEweb. n.d. "How to Value Ecosystem Services?" CEEweb for Biodiversity. Accessed November 13, 2019. https://www.ceeweb.org/work-areas/priority-areas/ecosystem-services/how-to-value-ecosystem-services

Center for Neighborhood Technology. 2018. "The Value of Green Infrastructure: A Guide to Recognizing Its Economic, Social and Environmental Benefits." Accessed October 14, 2019. https://www.cnt.org/publications/the-value-of-green-infrastructure-a-guide-to-recognizing-its-economic-environmental-and

Central Arizona–Phoenix LTER. 2019. "Central Arizona-Phoenix LTER Site Details." Accessed December 15, 2019. https://sustainability.asu.edu/caplter/

CEPF. 2019. "Biodiversity Hotspots Defined." Accessed August 6, 2019. https://www.cepf.net/our-work/biodiversity-hotspots/hotspots-defined

Cernansky, Rachel. 2017. "The Biodiversity Revolution." *Nature* 546: 23–24. Accessed August 6, 2019. https://www.nature.com/articles/546022a.pdf?origin=ppub

Cerra, Joshua F. 2015. "The YardWorks Project: Developing Urban Ecological Design Strategies for Residential Private Property." Manhattan, KS: CELA. Accessed October 23, 2019. https://thecela.org/wp-content/uploads/CERRA.pdf

Chiesura, Anna. 2004. "The Role of Urban Parks for the Sustainable City." *Landscape and Urban Planning* 68: 129–138.

Chiesura, Anna, and Joan Martinez-Alier. 2011. "How Much Is Urban Nature Worth? And for Whom?" In *The Routledge Handbook of Urban Ecology*, edited by Ian Douglas, David Goode, Mike Houck, and Rusong Wang, 93–95. New York: Routledge.

Chowdhury, Rinku Roy, Kelli Larson, Morgan Grove, Colin Polsky, Elizabeth Cook, Jeffery Onsted, and Laura Ogden. 2011. "A Multi-Scalar Approach to Theorizing Socio-Ecological Dynamics of Urban Residential Landscapes." *Cities and Environment* 4(1), Article 6.

Citizen Science Central. 2019. "What is Citizen Science and PPSR?" Accessed November 30, 2019. https://www.birds.cornell.edu/citizenscience/

Citizen Science Soil Collection Program. 2019. Accessed July 9, 2019. https://scistarter.org/drug-discovery-from-your-soil

City of Chicago. 2019. "City Hall's Rooftop Garden." Planning and Development Department. Accessed August 20, 2019. https://www.chicago.gov/city/en/depts/dgs/supp_info/city_hall_green_roof.html

City of Toronto. 2019. "Green Roof Overview." Accessed August 18, 2019. https://www.toronto.ca/city-government/planning-development/official-plan-guidelines/green-roofs/green-roof-overview/

Clausen, John C., Mark Hood, and Glenn S. Warner. 2006. "Low Impact Development Works! This Low Impact Development Demonstration Shows that Best Management Practices Do Reduce Storm Water Runoff." *Journal of Soil and Water Conservation* 61(2): 58A+.

Climate Positive Design. n.d. "Find Your Path to Being Climate Positive." Accessed October 14, 2019. https://www.atelierten.com/find-your-path-to-a-climate-positive-landscape/

Cobb, Mia. 2020. "The Poo Power! Global Challenge." Citizen Science Center. Accessed January 16, 2020. http://www.citizensciencecenter.com/poo-power-global-challenge/

Collado, Silvia, Jose A. Corraliza, Henk Staats, and Miguel Ruiz. 2015. "Effect of Frequency and Mode of Contact with Nature on Children's Self-Reported Ecological Behaviors." *Journal of Environmental Psychology* 41: 65–73. http://dx.doi.org/10.1016/j.jenvp.2014.11.001

Collier, Marcus. 2016. "A Role for Novel Ecosystems in the Anthropocene?" Accessed August 25, 2019. https://ugecviewpoints.wordpress.com/2016/09/20/a-role-for-novel-ecosystems-in-the-anthropocene/

Conniff, Richard. 2014. "Urban Nature: How to Foster Biodiversity in World's Cities." Accessed August 5, 2019. https://e360.yale.edu/features/urban_nature_how_to_foster_biodiversity_in_worlds_cities

Connolly, N.D.B. 2006. "Colored, Caribbean, and Condemned: Miami's Overtown District and the Cultural Expense of Progress, 1940–1970." *Caribbean Studies* 34(1): 3–60.

Conservation Gateway. n.d. "Payment for Ecosystem Services." Accessed July 29, 2019. https://www.conservationgateway.org/ConservationPractices/EcosystemServices/ValuationandPayments/PaymentforEcosystemServices/Pages/payment-ecosystem-service.aspx

Cook, Elizabeth M., Sharon J. Hall, and Kelli L. Larson. 2012. "Residential Landscapes as Social-Ecological Systems: A Synthesis of Multi-Scalar Interactions between People and Their Home Environment." *Urban Ecosystems* 15(1): 19–52.

Cooper, Bethany, Lin Crase, and Darryl Maybery. 2017. "Incorporating Amenity and Ecological Values of Urban Water into Planning Frameworks: Evidence from Melbourne, Australia." *Australian Journal of Environmental Management.* 24(1): 64–80.

Cooper, Caren B., Janis Dickinson, Tina Phillips, and Rick Bonney. 2007. "Citizen Science as a Tool for Conservation in Residential Ecosystems." *Ecology and Society* 12(2): 11. http://www.ecologyandsociety.org/vol12/iss2/art11/

Cortinovis, Chiara, and Davide Geneletti. 2018. "Ecosystem Services in Urban Plans: What Is There, and What Is Needed for Better Decisions." *Land Use Policy* 70: 298–310.

CWF (Canadian Wildlife Federation). 2021. "CWF Bioblitz: Help take Canada's Nature Selfie." Accessed March 11, 2021. http://cwf-fcf.org/en/explore/bioblitz/

Dahlman, LuAnn. 2019. "Eight Cities Slated to Run Urban Heat Island Mapping Campaigns in Summer 2019." Accessed August 17, 2019. https://cpo.noaa.gov/News/News-Article/ArtMID/6226/ArticleID/1704/Eight-cities-slated-to-run-Urban-Heat-Island-mapping-campaigns-in-summer-2019

Dai, Liping, Helena F.M.W. van Rijswick, Peter P. J. Driessen, and Andrea M. Keessen. 2018. "Governance of the Sponge City Programme in China with Wuhan as a Case Study." *International Journal of Water Resources Development* 34(4): 578–596.

Davis, Lucas W. 2013. *The Economic Cost of Global Fuel Subsidies*. Working Paper #19736. Cambridge, MA: National Bureau of Economic Research.

Defra. 2007. "An Introductory Guide to Valuing Ecosystem Services." Department of Environment, Food and Rural Affairs. Accessed November 14, 2019. https://ec.europa.eu/environment/nature/biodiversity/economics/pdf/valuing_ecosystems.pdf

———. 2009. "Safeguarding Our Soils, a Strategy for England." Accessed July 18, 2019. https://www.gov.uk/government/publications/safeguarding-our-soils-a-strategy-for-england

Defra Soils Policy Team. 2009. "Construction Code of Practice for the Sustainable Use of Soils on Construction Sites." Accessed July 8, 2019. https://assets.publishing.service.gov.uk/government/uploads/system/uploads/attachment_data/file/716510/pb13298-code-of-practice-090910.pdf

Del Tredici, Peter. 2014. "The Flora of the Future." *Places Journal* DOI: 10.22269/140417

Department of Labor. 2018. "Environmental Engineers." Bureau of Labor Statistics, Occupational Outlook Handbook." Accessed November 19, 2019. https://www.bls.gov/ooh/architecture-and-engineering/environmental-engineers.htm

Department of Political Science. 2019. "Environmental Politics and Policy Concentration." Accessed November 19, 2019. https://polisci.colostate.edu/majors-environmental-politics-policy/

Descoteau, D. 2017. "Victoria's Dockside Green Development Changes Hands." *Victoria News*, October 10, 2017.

De Sousa, Christopher A. 2004. "The Greening of Brownfields in American Cities." *Journal of Environmental Planning and Management* 47(4): 579–600.

———. 2006. "Unearthing the Benefits of Brownfield to Green Space Projects: An Examination of Project Use and Quality of Life Impacts." *Local Environment* 11(5): 577–600.

DEWHA. 2009. "Ecosystem Services: Key Concepts and Applications." Occasional Paper No 1, Department of the Environment, Water, Heritage and the Arts, Canberra. Accessed July 30, 2019. https://www.environment.gov.au/biodiversity/publications/ecosystem-services-key-concepts-and-applications

Dibner, K. A., and R. Pandya, editors. 2018. "Mapping the Landscape." *Learning Through Citizen Science: Enhancing Opportunities by Design*. Washington DC.: National Academies Press. Accessed November 27, 2019. https://www.ncbi.nlm.nih.gov/books/NBK535957/

Dickson, Dawn, and Richard J. Hobbs. 2017. "Cultural Ecosystem Services: Characteristics, Challenges and Lessons for Urban Green Space Research." *Ecosystem Services* 25: 179–194.

Diep, Francie. 2011. "Lawns vs. Crops in the Continental U.S." *Scienceline*. Accessed July 21, 2019. https://scienceline.org/2011/07/lawns-vs-crops-in-the-continental-u-s/

Dietz, Michael E., and John C. Clausen. 2008. "Stormwater Runoff and Export Changes with Development in a Traditional and Low Impact Subdivision." *Journal of Environmental Management* 87(4): 560–566.

Discover. 2019. "Top 18 Projects of 2018 on SciStarter-Globe at Night." *SciStarter*. Accessed September 28, 2019. https://scistarter.org/globe-at-night

Divisek, Jan, Milan Chytry, Brian Beckage, Nicholas J. Gotelli, Zdenka Lososova, Petr Pysek, David M. Richardson, and Jane Molofsky. 2018. "Similarity of Introduced Plant Species to Native Ones Facilitates Naturalization, but Differences Enhance Invasion Success." *Nature Communications* 9: 4631.

Dluhy, Milan, Keith Revell, and Sidney Wong. 2002. "Creating a Positive Future for a Minority Community: Transportation and Urban Renewal Politics in Miami." *Journal of Urban Affairs* 24(1): 75–95.

Dooling, Sarah. 2015. "Novel Landscapes: Challenges and Opportunities for Educating Future Ecological Designers and Restoration Practitioners." *Ecological Restoration* 33(1): 96–110.

Dorsey, Joseph W. 2010. "Lawn Control, Lawn Culture, and the Social Marketing of Sustainable Behaviors." *Ecopsychology* 2(2)(June): 91–103.

Douglas, Ian. 2011. "Suburban Mosaic of Houses, Roads, Gardens, and Mature Trees." In *The Routledge Handbook of Urban Ecology*, edited by Ian Douglas, David Goode, Mike Houck, and Rusong Wang, 264–273. New York: Routledge.

Douglas, Ian, and Philip James. 2015a. "Cities and Ecology." In *Urban Ecology, An Introduction*, 9–33. New York: Routledge.

———. 2015b. "The Urban Atmosphere, Weather, Climate, and Air Quality." In *Urban Ecology, An Introduction*, 75–100. New York: Routledge.

———. 2015c. "Adapting to Change." In *Urban Ecology, An Introduction*, 367–394. New York: Routledge.

———. 2015d. "Urban Hydrology." In *Urban Ecology, An Introduction*, 131–158. New York: Routledge.

———. 2015e. "Cities As Systems." In *Urban Ecology, An Introduction*, 56–72. New York: Routledge.

———. 2015f. "Urban Fauna." In *Urban Ecology, An Introduction*, 237–258. New York: Routledge.

———. 2015g. "The Role of Urban Ecology in Future Cities." In *Urban Ecology, An Introduction*, 395–419. New York: Routledge.

Downs, Anthony. 1999. "Some Realities about Sprawl and Urban Decline." *Housing Policy Debate* 10(4): 955–974.

———. 2005. "Smart Growth: Why We Discuss It More than We Do It." *Journal of the American Planning Association* 71(4): 367–380.

Downton, Paul, David Jones, Joshua Zeunert, and Phillip Barend Roös. 2017. *Creating Healthy Places: Railway Stations, Biophilic Design and the Metro Tunnel Project*. Accessed November 23, 2019. http://hdl.handle.net/10536/DRO/DU:30104224

Duffy, A. 2017. "New beginning for Dockside Green project in Vic West." *Times Colonist*, January 29, 2017.

Dunham-Jones, Ellen, and June Williamson. 2009. *Retrofitting Suburbia: Urban Design Solutions for Redesigning Suburbs*. Hoboken, NJ: John Wiley & Sons.

Dunn, Christopher, and Liam Heneghan. 2011. "Composition and Diversity of Urban Vegetation." In *Urban Ecology: Patterns, Processes, and Applications*, edited by Jari Niemelä, Jurgen Breuste, Thomas Elmqvist, Glenn Guntenspergen, Philip James, and Nancy E. McIntyre, 103–114. New York: Oxford University Press.

Dupont, Lien, Marc Antrop, and Veerle Van Eetvelde. 2014. "Eye-tracking Analysis in Landscape Perception Research: Influence of Photographic Properties and Land-scape Characteristics." *Landscape Research* 39(4): 417–432.

Earth Challenge 2020. n.d. Accessed July 31, 2019. https://www.earthday.org.

eBird. n.d. "About eBird." The Cornell Lab of Ornithology. Accessed January 21, 2018. https://ebird.org/home

Ecological Landscape Alliance. 2019. "Native Plant, Invasive Plant . . . What are the Differences? Why Does it Matter?" Accessed November 4, 2019. https://www.ecol-andscaping.org/native-and-invasive-plants/

EDD Maps. 2019. "Why Do Plants Become Invasive?" University of Georgia. Accessed November 4, 2019. https://www.eddmaps.org/about/why_plants_invade.cfm

Eldridge, Matt, Kimberly Burrowes, and Patrick Spauster. 2019. *Investing in Equitable Park Systems: Emerging Funding Strategies and Tools*. Washington, DC: Urban Institute.

Elmqvist, Thomas, Michail Fragkias, Julie Goodness, Burak Güneralp, Peter J. Marcotul-lio, Robert I. McDonald, Susan Parnell, Maria Schewenius, Marte Sendstad, Karen C. Seto, and Cathy Wilkinson, eds. 2013. *Urbanization, Biodiversity and Ecosystem Services: Challenges and Opportunities: A Global Assessment*. Dordrecht: Springer.

Elmqvist, Thomas, Charles L. Redman, Stephan Barthel, and Robert Costanza. 2013. "History of Urbanization and the Missing Ecology." In *Urbanization, Biodiversity and Ecosystem Services: Challenges and Opportunities: A Global Assessment*, edited by Thomas Elmqvist, Michail Fragkias, Julie Goodness, Burak Güneralp, Peter J. Marcotullio, Robert I. McDonald, Susan Parnell, Maria Schewenius, Marte Sendstad, Karen C. Seto, and Cathy Wilkinson, 13–30. Dordrecht: Springer.

Endreny, T., R. Santagata, A. Perna, C. De Stefano, R. F. Rallo, and S. Ulgiati. 2017. "Implementing and Managing Urban Forests: A Much Needed Conservation Strategy to Increase Ecosystem Services and Urban Wellbeing." *Ecological Modelling* 360: 328–335.

Environmental Science. 2019a. "What is an Environmental Lawyer?" Accessed November 19, 2019. https://www.environmentalscience.org/career/environmental-lawyer

———. 2019b. "What is an Environmental Economist?" Accessed November 19, 2019. https://www.environmentalscience.org/career/environmental-economist

———. 2019c. "What is an Environmental Sociologist?" Accessed November 19, 2019. https://www.environmentalscience.org/career/environmental-sociologist

EPA (United States Environmental Protection Agency). 2020. "Sustainable Marketplace: Greener Products and Services." Accessed January 16, 2020. https://www.epa.gov/greenerproducts

———. 2019a. US Environmental Protection Agency Archive. Accessed December 13, 2019. https://www.epa.gov/laws-regulations

———. 2019b. https://www.epa.gov/brownfields

———. 2019c. Accessed October 31, 2019. https://www.epa.gov/green-infrastructure/green-infrastructure-collaborative

———. 2015. "Green Infrastructure Statement of Intent." Accessed September 21, 2019. https://www.epa.gov/sites/default/files/2015-10/documents/gi_intentstatement.pdf

Erikson, Glenn Robert. 2016. "Strategies for Urban Ecology." *World Policy Journal*. Accessed November 5, 2019. https://worldpolicy.org/2016/12/28/strategies-for-urban-ecology/

ESA Annual Meeting. 2016. "Novel Ecosystems in Cities: Adaptation to Urban Conditions." https://eco.confex.com/eco/2016/webprogram/Session11678.html

European Environment Agency. 2016. "Waste: A Problem or a Resource?" Accessed December 3, 2019. https://www.eea.europa.eu/signals/signals-2014/articles/waste-a-problem-or-a-resource

Ewing, Reid, Rolf Pendall, and Don Chen. 2002. *Measuring Sprawl and Its Impact*. Washington, DC: Smart Growth America.

Ewing, Reid, Tom Schmid, Richard Killingsworth, Amy Zlot, and Stephen Raudenbush. 2003. "Relationship between Urban Sprawl and Physical Activity, Obesity, and Morbidity." *American Journal of Health Promotion* 18: 47–57.

Faith, Daniel P. 2016. "Biodiversity." *The Stanford Encyclopedia of Philosophy* (Fall 2019 Edition), edited by Edward N. Zalta. https://plato.stanford.edu/archives/sum2016/entries/biodiversity/

Feinberg, Daniel, and Mark Hostetler. 2013. "Conserving Urban Wildlife in the Face of Climate Change." WEC336. Accessed August 30, 2019. https://edis.ifas.ufl.edu/uw381

Feldkamp, Lisa. 2015. "Citizen Science that Speaks for Trees." Cool Green Science. Accessed July 15, 2019. https://blog.nature.org/science/2015/11/18/citizen-science-speaks-for-trees-i-tree-forests-urban/

FFL (Florida-Friendly Landscaping™ Program). 2017. "GI-BMP Training & Program Overview." Accessed October 3, 2019. https://ffl.ifas.ufl.edu/professionals/BMP_overview.htm

Fibershed. n.d. "Citizen Science Soil Sampling Protocol." Accessed July 8, 2019. http://fibershed.org/wp-content/uploads/2016/10/Soil-Sampling-Protocol.pdf

Fletcher, T. D., H. Andrieu, and P. Hamel. 2013. "Understanding, Management and Modelling of Urban Hydrology and its Consequences for Receiving Waters: A State of the Art." *Advances in Water Resources* 51: 261–279.

Ford, Larry R. 2003. *America's New Downtowns: Revitalization or Reinvention?* Baltimore: Johns Hopkins University Press.

Forman, Richard T. T. 1995. *Land Mosaics: The Ecology of Landscapes and Regions*. New York: Cambridge University Press.

Forman, Richard T. T., and Michel Godron. 1986. *Landscape Ecology*. New York: John Wiley & Sons.

Forman, Richard, T. T., and Jianguo Wu. 2016. "Where to Put the Next Billion People." *Nature* 537(September): 608–611.

Frank, Lawrence D., and Peter O. Engelke. 2001. "The Built Environment and Human Activity Patterns: Exploring the Impacts of Urban Form on Public Health." *Journal of Planning Literature* 16(2): 202–218.

Frank, Lawrence D., Peter O. Engelke, and Thomas L. Schmid. 2003. *Health and Community Design: The Impact of the Built Environment on Physical Activity*. Washington, DC: Island Press.

Franklin, Kimberly. 2007. "An Introduction to Reconciliation Ecology." Accessed December 8, 2019. http://tolweb.org/treehouses/?treehouse_id=4558

Freeman, Claire, Katharine J. M. Dickinson, Stefan Porter, and Yolanda van Heezik. 2012. "'My Garden is an Expression of Me': Exploring Householders' Relationships with Their Gardens." *Journal of Environmental Psychology* 32(2012): 135–143.

Freight Farms. 2020. "Introducing the Greenery." Accessed January 12, 2020. https://www.freightfarms.com/home#mission

Freyman, W. A., L. A. Masters, and S. Packard. 2016. "The Universal Floristic Quality Assessment (FAQ) Calculator: An Online Tool for Ecological Assessment and Monitoring." *Methods in Ecology and Evolution* 7(3): 380–383. Accessed October 14, 2019. https://universalfqa.org/about

Friedmann, John. 1987. *Planning in the Public Domain: From Knowledge to Action.* Princeton, NJ: Princeton University Press.

Frischenbruder, Marisa T. M., and Paulo Pellegrino. 2006. "Using Greenways to Reclaim Nature in Brazilian Cities." *Landscape and Urban Planning* 76(1–4): 67–78.

Fritz, Steffen, Linda See, Tyler Carlson, Mordechai Haklay, Jessie L. Oliver, Dilek Fraisl, Rosy Mondardini, et al. 2019. "Citizen Science and the United Nations Sustainable Development Goals." *Nature Sustainability* 2(10): 922–930.

Frumkin, Howard, ed. 2010. *Environmental Health: From Global to Local.* 2nd ed. San Francisco: John Wiley & Sons.

Frumkin, Howard, Lawrence D. Frank, and Richard Jackson. 2004. *Urban Sprawl and Public Health: Designing, Planning, and Building for Healthy Communities.* Washington, DC: Island Press.

Fry, G., M. S. Tveit, A. Ode, and M. D. Velarde. 2009. "The Ecology of Visual Landscapes: Exploring the Conceptual Common Ground of Visual and Ecological Landscape Indicators." *Ecological Indicators* 9(2009): 933–947.

Gallo, Travis, and Mason Fidino. 2018. "Biodiversity: Making Wildlife Welcome in Urban Areas." Accessed July 24, 2019. https://elifesciences.org/articles/41348

Garden Research. 2018. "Gardening Reaches an All Time High." Accessed October 18, 2019. https://www.globenewswire.com/news-release/2018/04/18/1480986/0/en/Gardening-Reaches-an-All-Time-High.html

Gardens by the Bay. n.d. "Our Story." Accessed November 19, 2019. https://www.gardensbythebay.com.sg/en/the-gardens/our-story/introduction.html

Garreau, Joel. 1991. *Edge City: Life on the New Frontier.* New York: Doubleday.

Gaston, K. J., Z. G. Davies, and J. L. Edmondson. 2010. "Urban Environments and Ecosystem Functions." In *Urban Ecology*, edited by K. J. Gaston, 35–52. Cambridge: Cambridge University Press.

Gehl Institute. 2017. "Public Life Tools." Accessed October 14, 2019. https://www.Gehlinstitute.org/tools/

Gerber, Leah. 2010. "Conservation Biology." Accessed August 7, 2019. https://www.nature.com/scitable/knowledge/library/conservation-biology-16089256

Gill, Jacquelyn. 2016. "Ecological Novelty in the Anthropocene: Are Novel Communities Novel Ecosystems?" Accessed August 25, 2019. https://eco.confex.com/eco/2016/webprogram/Session11727.html

Gillham, Oliver. 2002. *The Limitless City: A Primer on the Urban Sprawl Debate*. Washington, DC: Island Press.

Girot, Christophe. 2016. *The Course of Landscape Architecture*. London: Thames & Hudson.

Gobster, Paul, H. 2007. "Urban Park Restoration and the 'Museumification' of Nature." *Nature and Culture* 2(2): 95–114.

Gobster, Paul H., Joan I. Nassauer, Terry C. Daniel, and Gary Fry. 2007. "The Shared Landscape: What Does Aesthetics Have to do with Ecology?" *Landscape Ecology* 22: 959–972.

Goode, David. 2015. "London: A National Park City." Accessed December 8, 2019. https://www.thenatureofcities.com/2015/08/16/london-a-national-park-city/

Grabow, Steven H. 2015. "Principles and Practice of Community Placemaking." G4083, University of Wisconsin Extension. Accessed October 27, 2019. https://blogs.extension.wisc.edu/community/files/2014/08/Placemaking-Document-Principles-and-Practice-of-Community-Placemaking-updated-5-14-14.pdf

Graham, Rod. 2004. "Ecologist Urges Sharing Land with Other Species to Foster Biodiversity." *Johns Hopkins Bloomberg School of Public Health News*, March 22, 2004. Accessed December 8, 2019. https://www.jhsph.edu/news/stories/2004/reconciliation.html

Grassmick, David E. 2002. "Minding the Neighbor's Business: Just How Far Can Condominium Owners' Associations Go in Deciding Who Can Move into the Building." *University of Illinois Law Review* 2002(1): 185–214.

Green, Jemma, David Martin, and Meagan Cojocar. 2018. "The Untouched Market: Distributed Renewable Energy in Multitenanted Buildings and Communities." In *Urban Energy Transition*, edited by Peter Droege, 401–418. Elsevier.

Green, Jemma, and Peter Newman. 2017. "Citizen Utilities: The Emerging Power Paradigm." *Energy Policy* 105: 283–293.

Greenberg, Michael, Karen Lowrie, Henry Mayer, K. Tyler Miller, and Laura Solitare. 2001. "Brownfield Redevelopment as a Smart Growth Option in the United States." *The Environmentalist* 21(2): 129–143.

Greencity Solutions. n.d. "The Solution to Quantifiably Improve City Air." Accessed January 16, 2020. https://greencitysolutions.de/en/solutions/

Green Facts. n.d. "Ecosystem Services." Accessed July 29, 2019. https://www.greenfacts.org/glossary/def/ecosystem-services.htm

Green Seal. 2019. Accessed August 21, 2019. https://www.greenseal.org

Grimmond, C.S.B. 2011. "Climate of Cities." In *The Routledge Handbook of Urban Ecology*, edited by Ian Douglas, David Goode, Mike Houck, and Rusong Wang, 103–119. New York: Routledge.

Groundplay SF. n.d. "A Look at the Human Impact of Parklets and the People Who Make Them." Accessed January 12, 2020. https://groundplaysf.org/wp-content/parklet-impact-study/

GSA. n.d. "Federal Crowdsourcing and Citizen Science Catalog." *Catalog Citizen Science.gov*. Accessed January 19, 2020. https://www.citizenscience.gov/catalog/45/#).

The Guardian. 2019. "Inside China's Leading 'Sponge City': Wuhan's War with Water."

Accessed November 1, 2019. https://www.theguardian.com/cities/2019/jan/23/inside-chinas-leading-sponge-city-wuhans-war-with-water

Gunderson, Lance H., C. S. Holling, and Stephen S. Light, eds. 1995. *Barriers and Bridges to the Renewal of Ecosystems and Institutions.* New York: Columbia University Press.

Hall, Peter G. 2002. *Cities of Tomorrow: An Intellectual History of Urban Planning and Design in the Twentieth Century,* 3rd ed. Malden, MA: Blackwell Publishers.

Handy, Susan L., Marlon G. Boarnet, Reid Ewing, and Richard E. Killingsworth. 2002. "How the Built Environment Affects Physical Activity: Views from Urban Planning." *American Journal of Preventive Medicine* 23(2S): 64–73.

Hanks, Dallas, and Ann Lewandowski. 2003. "Protecting Urban Soil Quality: Examples for Landscape Codes and Specifications." Accessed July 15, 2019. https://www.nrcs.usda.gov/Internet/FSE_DOCUMENTS/nrcs142p2_053275.pdf.

Hansen, Gail. 2010. "Basic Principles of Landscape Design." CIR536 Environmental Horticulture Department. https://edis.ifas.ufl.edu/mg086

Hansen, Gail, and Shangchun Hu. 2013. "Florida-Friendly Plants for Stormwater Pond Shorelines." Publication #ENH1215. https://edis.ifas.ufl.edu/ep476

Harnick, Peter. 2000. *Inside City Parks.* Washington, DC: Trust for Public Land.

Haughton, Graham. 1999. "Environmental Justice and the Sustainable City." *Journal of Education and Planning Research* 18 (3): 233–243.

Havens, Karl. 2018. "The Future of Harmful Algal Blooms in Florida Inland and Coastal Waters." Document TP-231. https://edis.ifas.ufl.edu/sg153

Haywood, Benjamin K. 2014. "A 'Sense of Place' in Public Participation in Scientific Research." *Science Education* 98(1): 64–83.

Heerwagen, Judith H. 2000. "Do Green Buildings Enhance the Well Being of Workers? Yes." *Environmental Design and Construction* 3(July/August): 24–29.

———2002. "Sustainable Design Can Be an Asset to the Bottom Line." *Environmental Design and Construction* 5(4): 35–39.

Heise, Ursula. 2019. "Imagining Future Cities in an Age of Ecological Change." The Nature of Cities. Accessed November 10, 2019. https://www.thenatureofcities.com/2019/05/27/imagining-future-cities-in-an-age-of-ecological-change/

Helfand, Gloria, E., Joon Sik Park, Joan I. Nassauer, and Sandra Kosek. 2006. "The Economics of Native Plants in Residential Landscape Designs." *Landscape and Urban Planning* 78: 229–240.

Hestnes, Anne Grete, and Nancy Lea Eik-Nes, eds. 2017. *Zero Emission Buildings.* Bergen: Fagbokforlaget.

High Line. 2020. "History." Accessed January 11, 2020. https://www.thehighline.org/history/

Hillsborough County. 2016. "Imagine 2040: City of Tampa Florida, Tampa Comprehensive Plan. Vision, People, Places, Natural Spaces Governance and Implementation," 239–290. Accessed September 24, 2019. http://www.planhillsborough.org/wp-content/uploads/2014/12/Tampa-2040-Comp-Plan-September2015.pdf

Historic England. 2020. "Historic Landscape Characterization." Accessed October 9, 2019. https://historicengland.org.uk/research/methods/characterisation-2/historic-landscape-characterisation/

Hobbs, Richard J., Eric Higgs, and James A. Harris. 2009. "Novel Ecosystems: Implications for Conservation and Restoration." *Trends in Ecology and Evolution* 24(11): 599–605

Hoff, Mary. 2018. "As Insect Populations Decline, Scientists Are Trying to Understand Why." Accessed July 25, 2019. https://www.scientificamerican.com/article/as-insect-populations-decline-scientists-are-trying-to-understand-why/

Hoffman, Jeremy. 2018. "Where Do We Need Shade? Mapping Urban Heat Islands in Richmond, Virginia." Accessed August 17, 2019. https://toolkit.climate.gov/case-studies/where-do-we-need-shade-mapping-urban-heat-islands-richmond-virginia

Hollander, Justin B. 2009. *Polluted and Dangerous: America's Worst Abandoned Properties and What Can Be Done About Them.* Burlington: University of Vermont Press, imprint of University Press of New England.

Holzman, David C. 2012. "Accounting for Nature's Benefits: The Dollar Value of Ecosystem Services." *Environmental Health Perspectives* 120(4): a152–a157.

Howland, Jon. 2012. "Clothesline Bans Void in 19 States." *Sightline Institute.* Accessed December 11, 2019. https://www.sightline.org/2012/02/21/clothesline-bans-void-in-19-states/

Hoyle, Helen, James D. Hitchmough, and Anna Jorgensen. 2017. "Attractive, Climate-adapted and Sustainable? Public Perception of Non-native Planting in the Designed Urban Landscape." *Landscape and Urban Planting* 164: 49–63.

Hu, Shangchun, and Ann Keeley. 2013. "Sustainable Urban Waters: Opportunities to Integrate Environmental Protection in Multi-Objective Projects." National Research Council, U.S. Environmental Protection Agency. Office of Research and Development, National Risk Management.

Huber, Jennifer. 2018. "Citizen Science Research Investigates Neighborhoods' Effects on Well-Being." Accessed December 21, 2019. https://scopeblog.stanford.edu/2018/06/26/citizen-science-research-investigates-neighborhoods-effects-on-well-being/

Hummel, Susan Stevens. 2016. "E. N. Anderson: Caring for Place: Ecology, Ideology, and Emotion in Traditional Landscape Management." *Agriculture and Human Values* 33: 495–496.

iNaturalist. 2017. Accessed July 23, 2019. https://www.inaturalist.org

Ingham, Lucy. 2014. "Plant Power: The New Technology Turning Green Roofs into Living Power Plants." Accessed July 16, 2019. https://www.factor-tech.com/green-energy/1569-plant-power-the-new-technology-turning-green-roofs-into-living-power-plants/

InVEST. 2016. "Integrated Valuation of Ecosystem Services and Tradeoffs." Accessed October 14, 2019. https://naturalcapitalproject.stanford.edu/software/invest

IPCC (Intergovernmental Panel on Climate Change). 2019. Accessed December 4, 2019. https://www.ipcc.ch

IPM. 2019. "What is Integrated Pest Management?" IPM Institute of North America. Accessed November 5, 2019. https://ipminstitute.org/what-is-integrated-pest-management/

Irwin, Aisling. 2018. "No PhDs Needed: How Citizen Science is Transforming Research." Accessed November 30, 2019. https://www.nature.com/articles/d41586-018-07106-5

Jarvis, Peter, J. 2011. "Urban Animal Ecology." In *The Routledge Handbook of Urban Ecol-*

ogy, edited by Ian Douglas, David Good, Mike Houck, and Rusong Wang, 352–360. New York: Routledge.

Jia, Haifeng, Zheng Wang, Xiaoyue Zhen, Mike Clar, and Shaw L. Yu. 2017. "China's Sponge City Construction: A Discussion on Technical Approaches." *Frontiers of Environmental Science & Engineering* 11(4): 18.

Kaplan, Rachel. 2001. "The Nature of the View from Home: Psychological Benefits." *Environment and Behavior* 33(4): 507–542.

Kaplan, Rachel, and Stephen Kaplan. 1989. *The Experience of Nature: A Psychological Perspective*. Cambridge: Cambridge University Press.

Kaplan, Stephen. 1987. "Aesthetics, Affect, and Cognition: Environmental Preference from an Evolutionary Perspective." *Environment and Behavior* 19(13): 3–32.

Kassab, Beth. 2016. "Lessons from HOA Lawsuit over Florida-Friendly Grass." Accessed October 4, 2019. https://www.orlandosentinel.com/opinion/os-hoa-fight-over-grass-beth-kassab-20160908-column.html

Katz, Bruce, and Jeremy Nowak. 2018. "How Cities Can Fund the Future." Accessed October 14, 2018. https://www.citylab.com/equity/2018/01/how-cities-can-fund-the-future/551630/

Kaufman, A. J., and V. I. Lohr. 2004. "Does Plant Color Affect Emotional and Physiological Responses to Landscapes?" *Acta Horticulturae* 639: 229–233.

Kaymaz, Isil. 2013. "Urban Landscapes and Identity: Advances in Landscape Architecture." Murat Ozyavuz, IntechOpen. Accessed October 28, 2019. https://www.intechopen.com/books/advances-in-landscape-architecture/urban-landscapes-and-identity

Kellert, Stephen R. 2012. *Birthright: People and Nature in the Modern World*. New Haven, CT: Yale University Press.

———2018. *Nature by Design: The Practice of Biophilic Design*. New Haven, CT: Yale University Press.

Kellert, Stephen R., and Elizabeth F. Calabrese. 2015. *The Practice of Biophilic Design*. www.biophilic-design.com

Kellert, Stephen R., Judith H. Heerwagen, and Martin L. Mador, eds. 2008. *Biophilic Design: The Theory, Science, and Practice of Bringing Buildings to Life*. Hoboken, NJ: John Wiley & Sons.

Kellert, Stephen R., and Edward O. Wilson, eds. 1993. *The Biophilia Hypothesis*. Washington, DC: Island Press.

Kendal, Dave, Kathryn J. H. Williams, and Nicholas S. G. Williams. 2012. "Plant Traits Link People's Plant Preferences to the Composition of Their Gardens." *Landscape and Urban Planning* 105: 34–42.

Kibert, Charles J., ed. 1999. *Reshaping the Built Environment: Ecology, Ethics, and Economics*. Washington, DC: Island Press.

Kimbrough, Liz. 2016. "Cities as Novel Ecosystems: Adaptations to Urban Conditions." Accessed August 6, 2019. https://theplosblog.plos.org/2016/08/novel-ecosystem-in-cities-adaptions-to-urban-conditions/

Kimmelman, Michael. 2011. "In a Bronx Complex, Doing Good Mixes with Looking Good." *New York Times*, September 26, 2011. Accessed December 16, 2019. https://www.nytimes.com/2011/09/26/arts/design/via-verde-in-south-bronx-rewrites-low-income-housing-rules.html

Kobilinsky, Dana. 2018. "City Hawks Flock to Feeders to Find Prey." Accessed July 25, 2019. https://wildlife.org/city-hawks-flock-to-feeders-to-find-prey/

———. 2019. "Citizen Scientists Point to Urban Carnivore Hotspots." Accessed July 23, 2019. https://wildlife.org/citizen-scientists-point-to-urban-carnivore-hotspots/

Kollmuss, Anja, and Julian Agyeman. 2010. "Mind the Gap: Why Do People Act Environmentally and What Are the Barriers to Pro-Environmental Behavior?" *Environmental Education Research* 8(3): 239–260.

Kovacs, Zsuzsi I., Carri J. LeRoy, Dylan G. Fischer, Sandra Lubarsky, and William Burke. 2006. "How Do Aesthetics Affect Our Ecology?" *Aesthetics and Ecology* 10: 61–65.

Kowarik, Ingo. 2011. "Novel Urban Ecosystems, Biodiversity, and Conservation." *Environmental Pollution* 159: 1974–1983.

Krueckeberg, Donald A. 1983. *Introduction to Planning History in the United States.* New Brunswick, NJ: Center for Urban Policy Research.

Kruuse, A. 2011. "GRaBS Expert Paper 6: The Green Space Factor and the Green Points System." In GRaBS, ed., *The GRaBS Project.* London: Town and Country Planning Association & GRaBS.

Kythreotis, Andrew P., Chrystal Mantyka-Pringle, Theresa G. Mercer, Lorraine E. Whitmarsh, Adam Corner, Jouni Pavvola, Chris Chambers, Byron A. Miller, and Noel Castree. 2019. "Citizen Social Science for More Integrative and Effective Climate Action: A Science-Policy Perspective." *Frontiers in Environmental Science.* Accessed November 1, 2019. https://www.frontiersin.org/articles/10.3389/fenvs.2019.00010/full

LaCroix, Anthony. 2006. "Overview of Feral Cat Population Control." Accessed July 24, 2019. https://www.animallaw.info/article/overview-feral-cat-population-control

LAF. 2018. "Evaluating Landscape Performance, A Guidebook for Metrics and Methods Selection." Accessed October 15, 2019. https://www.landscapeperformance.org/guide-to-evaluate-performance

La Follette, Cameron. 2019. "Rights of Nature: The New Paradigm." Accessed November 11, 2019. https://news.aag.org/2019/03/rights-of-nature-the-new-paradigm

LaFrance, Adrienne. 2015. "When You give a Tree an Email Address." Accessed November 22, 2019. https://www.theatlantic.com/technology/archive/2015/07/when-you-give-a-tree-an-email-address/398210/

Lam, Kristin. 2019. "Georgia Gains Major Court Victory over Florida in Decades-Long Dispute over Apalachicola Bay Water Use." *USA Today.* Accessed December 13, 2019. https://www.usatoday.com/story/news/nation/2019/12/12/georgia-florida-apalachicola-water-use/4414478002/

Landscape Performance Series. 2018. "Case Study Briefs, Renaissance Park." Accessed October 14, 2018. https://www.landscapeperformance.org/case-study-briefs/renaissance-park

Larson, Kelli L., and Jaleila Brumand. 2014. "Paradoxes in Landscape Management and Water Conservation: Examining Neighborhood Norms and Institutional Forces." *Cities and the Environment* 7, Iss. 1, Article 6. http://digitalcommons.lmu.edu/cate/vol7/iss1/6

Lathrop, Janet. 2014. "Urban Bird, Plant Species More Diverse Worldwide than Expected." Accessed August 6, 2019. https://www.umass.edu/archivenewsoffice/article/urban-bird-plant-species-more-diverse

Law, Alan, Martin J. Gaywood, Kevin C. Jones, Paul Ramsay, and Nigel J. Willby. 2017. "Using Ecosystem Engineers as Tools in Habitat Restoration and Rewilding: Beaver and Wetlands." *Science of the Total Environment* 605–606: 1021–1030.

Ledsom, Alex. 2019. "What Is London's National Park City Status, and Which Other Cities Will Follow?" Accessed December 8, 2019. https://www.forbes.com/sites/alexledsom/2019/07/30/what-is-londons-national-city-park-status-and-which-other-cities-will-follow/#6d99f7347a7a

Lee, Chanam, and Anne Vernez Moudon. 2004. "Physical Activity and Environment Research in the Health Field: Implications for Urban and Transportation Planning Practice and Research." *Journal of Planning Literature* 19(2): 147–181.

Lepczyk, Christopher A., Myla F. J. Aronson, Karl L. Evans, Mark A. Goddard, Susannah B. Lerman, and J. Scott Macivor. 2017. "Biodiversity in the City: Fundamental Questions for Understanding the Ecology of Urban Green Spaces for Biodiversity Conservation." *Bioscience* 6(9): 799–807. https://academic.oup.com/bioscience/article/67/9/799/4056044

Levine, Jeremy R. 2013. "When Environmental and Social Policy Converge: The Case of Boston's Fairmount Line." In *Environmental Policy Is Social Policy—Social Policy is Environmental Policy*, edited by Isidor Wallimann, 153–163. New York: Springer.

Ley, Shaun. 2015. "The Melbourne Treemail Phenomenon." Accessed July 17, 2019. https://www.bbc.com/news/magazine-33560182

Li, Qian, Feng Wang, Yang Yu, Zhengce Huang, Mantao Li, and Yuntao Guan. 2019. "Comprehensive Performance Evaluation of LID Practices for the Sponge City Construction: A Case Study in Guangxi, China." *Journal of Environmental Management* 231: 10–20.

LID Center. 2019. "Low Impact Development Center." Accessed October 31, 2019. https://lowimpactdevelopment.org/projects/

Lieber, John. 2018. "Urban Ecology: A Bright Future for Sustainable Cities." The Revelator. Accessed November 20, 2019. https://therevelator.org/urban-ecology-sustainable-cities/

Ligenfelter, Dwight. 2009. "Introduction to Weeds: What Are Weeds and Why Do We Care?" Accessed July 14, 2019. https://extension.psu.edu/introduction-to-weeds-what-are-weeds-and-why-do-we-care

Lin, Xiaohu, Jie Ren, Jingcheng Xu, Tao Zheng, Wei Cheng, Junlian Qiao, Juwen Huang, and Guangming Li. 2018. "Prediction of Life Cycle Carbon Emissions of Sponge City Projects: A Case Study in Shanghai, China." *Sustainability* 10(11): 16.

Livingstone, Aislin. 2017. "The Use of Citizen Science in the Landscape Design Process: Opportunities for Monitoring and Evaluation." Thesis project presented to the University of Guelph. Accessed October 2, 2019. https://pdfs.semanticscholar.org/4c81/617c6c41294805e8589f148d691961530fc7.pdf

Loder, A. 2014. "'There's a Meadow Outside My Workplace': A Phenomenological Exploration of Aesthetics and Green Roofs in Chicago and Toronto." *Landscape and Urban Planning* 126: 94–106.

Loukaitou-Sideris, Anastasia. 2018. *SMART Parks: A Toolkit*. Los Angeles: UCLA Luskin Center for Innovation.

Louv, Richard. 2010. *Last Child in the Woods: Saving Our Children from Nature-Deficit Disorder*. Rev. ed. London: Atlantic.

MacDonagh, Peter. 2016. "Can Dirt Become Soil Again?" Accessed July 20, 2019. www.deeproot.com/blog/blog-entries/can-dirt-become-soil-again

Macedo, Joseli. 2013. "Planning a Sustainable City: The Making of Curitiba, Brazil." *Journal of Planning History* 12(4): 333–352.

Macedo, Joseli, and Mônica A. Haddad. 2016. "Equitable Distribution of Open Space: Using Spatial Analysis to Evaluate Parks in Curitiba, Brazil." *Environment & Planning B: Planning and Design* 43(6): 1096–1117.

Maddox, David. 2013. "The Cities We Want: Resilient, Sustainable, and Livable." The Nature of Cities. Accessed December 1, 2019. https://www.thenatureofcities.com/2013/05/08/the-cities-we-want-resilient-sustainable-and-livable/

MAFF. 2000. "Good Practice Guide for Handling Soils." Accessed July 8, 2019. https://webarchive.nationalarchives.gov.uk/20090317221756/http://www.defra.gov.uk/farm/environment/land-use/soilguid/index.htm

Mancebo, François. 2017. "Future Cities Live Underground—And That's Not a Pile of Schist." Accessed December 4, 2019. https://www.thenatureofcities.com/2017/01/22/future-cities-live-underground-thats-not-pile-schist/

Manning-Hudson, Laura. 2012. "Lawsuit by Neighborhood Association Against Homeowner for Water-Conserving Yard Will Test Florida's Xeriscaping Law." Accessed October 4, 2019. https://www.floridahoalawyerblog.com/lawsuit-by-neighborhood-associ/

Marble, Chris, and Andrew Koeser. 2018. "Improving Weed Control in Landscape Planting Beds." ENH1262. Accessed July 16, 2019. https://edis.ifas.ufl.edu/ep523

Marcotuillo, Peter. 2011. "Urban Soils." In *The Routledge Handbook of Urban Ecology*, edited by Ian Douglas, David Good, Mike Houck, and Rusong Wang, 164–181. New York: Routledge.

Martel, Yann. n.d. "Quotable Quote from Life of Pi." Eleventh Annual Goodreads Choice Awards. Accessed November 25, 2019. https://www.goodreads.com/quotes/259216-if-you-took-the-city-of-tokyo-and-turned-it

Matisoff, D., and D. Noonan. 2012. "Managing Contested Greenspace: Neighborhood Commons and the Rise of Dog Parks." *International Journal of the Commons* 6(1): 28–51.

Matysek, Kate. 2004. "Theory and Planning for Urban Biosphere Reserves: An Australian Example." *Leading Edge 2004: The Working Biosphere*.

Maulan, Suhardi, Mustafa Kamal Mohd. Shariff, and Patrick Miller. 2006. "Landscape Preference and Human Well-Being." *International Journal on Sustainable Tropical Design Research & Practice* 1(1): 25–32.

McCance, Erin C., Daniel J. Decker, Anne M. Colturi, Richard K. Baydack, William F. Siemer, Paul D. Curtis, and Thomas Eason. 2017. "Importance of Urban Wildlife Management in the United States and Canada." *Mammal Study* 42(1): 1–16. https://doi.org/10.3106/041.042.0108.

McDonald, David, Howard Stenn, and Jim Berger. 2011. "The Science and Practice of Sustainable Sites: Practical Implementation of Soil Protection & Restoration." Accessed July 19, 2019 https://depts.washington.edu/uwbg/docs/sites/Sites_Soil_Mc-Donald_Stenn_Berger.pdf

McDonough, William, and Michael Braungart. 2002. *Cradle to Cradle: Remaking the Way We Make Things*. New York: North Point Press.

McKinley, Duncan C., Abe J. Miller-Rushing, Heidi L. Ballard, Rick Bonney, Hutch Brown, Susan C. Cook-Patton, Daniel M. Evans, Rebecca A. French, Julia K. Parrish, Tina B. Phillips, et al. 2017. "Citizen Science Can Improve Conservation Science, Natural Resource Management, and Environmental Protection." *Biological Conservation* 208: 15–28.

McKinstry, Mark C., Paul Caffrey, and Stanley H. Anderson. 2001. "The Importance of Beaver to Wetland Habitats and Waterfowl in Wyoming." *Journal of the American Water Resources Association* 37(6): 1571–1577.

McPhearson, Timon, David Maddox, Bram Gunther, and David Bragdon. 2013. "Local Assessment of New York City: Biodiversity, Green Space, and Ecosystem Services." In *Urbanization, Biodiversity and Ecosystem Services: Challenges and Opportunities, a Global Assessment*, edited by Thomas Elmqvist, Michail Fragkias, Julie Goodness, Burak Güneralp, Peter J. Marcotullio, Robert I. McDonald, Susan Parnell, Maria Schewenius, Marte Sendstad, Karen C. Seto, and Cathy Wilkinson, 355–383. Dordrecht: Springer.

McPhearson, Timon, Steward T. A. Pickett, Nancy B. Grimm, Jari Niemelä, Marina Alberti, Thomas Elmqvist, Christine Weber, Dagmar Haase, Jurgen Breuste, and Salman Qureshi. 2016. "Advancing Urban Ecology toward a Science of Cities." *BioScience* 66(3): 198–212.

McPhearson, Timon, Susan Parnell, David Simon, Owen Gaffney, Thomas Elmqvist, Xuemei Bai, Debra Roberts, et al. 2016. "Scientists Must Have a Say In The Future of Cities." *Nature* 538 (October): 165–166.

McWhinney, James. 2019. "Water Investments: How to Invest in Water." Accessed January 20, 2020. https://www.investopedia.com/articles/06/water.asp

Medina, Jon. 2018. "Vision Trumps All Other Senses." *Brain Rules*. Accessed January 14, 2020. http://www.brainrules.net/vision?scene=

Mei, Chao, Jiahong Liu, Hao Wang, Zhiyong Yang, Xiangyi Ding, and Weiwei Shao. 2018. "Integrated Assessments of Green Infrastructure for Flood Mitigation to Support Robust Decision-Making for Sponge City Construction in an Urbanized Watershed." *Science of the Total Environment* 639: 1394–1407.

Middleton, Julie V. 2001. "The Stream Doctor Project: Community-Driven Stream Restoration." *BioScience* 51(4): 293–296.

Millennium Ecosystem Assessment. 2003. "Ecosystems and Human Well-being: A Framework for Assessment." Accessed July 29, 2019. https://www.millenniumassessment.org/en/Framework.html

———. 2005. *Ecosystems and Human Well-being*. Washington, DC: Island Press. https://pdf.wri.org/ecosystems_human_wellbeing.pdf.

Miller, Jim. 2016. "What's Wrong with Novel Ecosystems, Really?" Accessed August 25, 2019. http://millerlab.nres.illinois.edu/pdfs/Miller_2016_What%27s%20wrong%20with%20novel%20ecosystems,%20really.pdf

Minogue, Kristen. 2012. "What the Plantation Owners Left Behind." Smithsonian Environmental Research Center. Accessed November 6, 2019. https://sercblog.si.edu/what-the-plantation-owners-left-behind/

Mitra, Smita. 2017. "Novel Ecosystems: How Our Cities Are Changing Nature." Accessed August 25, 2019. http://citizenmatters.in/novel-ecosystems-urban-biodiversity-trees-nature-4404

Mize, Alison. 2015. "Expanding the Reach of Environmental Research with Citizen Science." Accessed August 27, 2019. https://www.esa.org/blog/2015/09/30/expanding-the-reach-of-environmental-research-with-citizen-science/

Modi, Sheetal. 2014. "Uncovering 'Ancient Lives' With Citizen Science." Accessed October 8, 2019. https://blog.scistarter.com/2014/01/ancient-lives/

Moebius-Clune, Bianca, Daniel J. Moebius-Clune, B. K. Gugiono, O. J. Idowu, Robert R. Schindelbeck, Aaron J. Ristow, Harold M. van Es, et al. 2016. "Comprehensive Assessment of Soil Health—The Cornell Framework." Edition 3.2, Cornell University, Geneva, NY. Accessed July 9, 2019. www.css.cornell.edu/extension/soil-health/manual-print.pdf

Mohawk. 2019. "Mohawk Sustainability." Accessed January 19, 2020. https://www.mohawkcollege.ca/sustainability

Mokori, A. S., and C. E. Cloete. 2016. "Ethical Perspectives on the Environmental Impact of Property Development." *HTS Teologiese Studies/Theological Studies* 72(3): a3209.

Momol, Esen, Jane Tolbert, Marina D'Abreau, Terril Nell, Gail Hansen, Gary Knox, Michael Thomas, et al. 2014. "Questions and Answers: 2009 Florida-Friendly Landscaping Legislation." Accessed October 3, 2019. https://edis.ifas.ufl.edu/pdffiles/EP/EP44000.pdf

Monteiro, José A. 2017. "Ecosystem Services from Turfgrass Landscapes." *Urban Forestry & Urban Greening* 26(August): 151–157.

Mooney, Chris. 2015. "Americans are Judging Their Neighbors' Lawns—With Surprising Environmental Consequences." Accessed September 23, 2019. https://www.washingtonpost.com/news/energy-environment/wp/2015/03/11/forget-what-your-neighbors-think-stop-dousing-your-lawn-with-so-much-fertilizer/

Moore, Keith Diaz. 2001. "The Scientist, the Social Activist, the Practitioner and the Cleric: Pedagogical Exploration towards a Pedagogy of Practice." *Journal of Architectural and Planning Research* 18(1): 59–79.

Morse, Nathaniel B., Paul A. Pellissier, Elisabeth N. Cianciola, Richard L. Brereton, Marleigh M. Sullivan, Nicholas K. Shonka, Tessa B. Wheeler, and William H. McDowell. 2014. "Novel Ecosystems in the Anthropocene: A Revision of the Novel Ecosystem Concept for Pragmatic Applications." *Ecology and Society* 19(2): 12.

Mortice, Zach, Kim O'Connell, Haniya Rae, and Timothy A. Schuler. 2019a. "The 2019 ASLA Professional Awards, Site Commissioning: Proving Triple-Bottom-Line Landscape Performance at a National Scale." *Landscape Architecture Magazine* 109(10): 118–190.

Mortice, Zach, Kim O'Connell, Haniya Rae, and Timothy A. Schuler. 2019b. "The 2019 ASLA Professional Awards, Chulalongkorn University Centenary Park." *Landscape Architecture Magazine* 109(10): 170–171.

Mott, Maryann. 2004. "U.S. Faces Growing Feral Cat Problem." Accessed July 24, 2019. https://www.nationalgeographic.com/animals/2004/09/feral-cat-problem/

Müller, Felix, Rudolf de Groot, and Louise Willemen. 2010. "Ecosystem Services at the Landscape Scale: the Need for Integrative Approaches." *Landscape Online* 23: 1–11.

MVRDV. n.d. "Seoullo 7017 Skygarden." Accessed December 15, 2019. https://mvrdv.nl/projects/208/seoullo-7017-skygarden

NACWA. 2009. "Clean Water Act Enforcement: Challenges and Opportunities in the 21st Century." White Paper. Accessed September 21, 2019. https://www.nacwa.org/docs/default-source/news-publications/White-Papers/2009-12-14enforcm-wp.pdf?sfvrsn=4

NASA. 2018. "New Citizen Science Projects Funded for Earth Studies." Accessed September 1, 2019. https://www.nasa.gov/feature/new-citizen-science-projects-funded-for-earth-studies

Nassauer, Joan I. 1988. "The Aesthetics of Horticulture: Neatness as a Form of Care." *HortScience* 23(6): 973–977

National Gardening Association. n.d. "National Gardening Association Survey 1999." Accessed July 8, 2019. https://gardenresearch.com

National Geographic. 2019. "Cities of the Future." In *Cities: Ideas for a Brighter Future, National Geographic*, April Special Issue.

———. n.d. "BioBlitz." Accessed March 11, 2021. https://www.nationalgeographic.org/encyclopedia/bioblitz/

National Recreation and Park Association. 2019. "Revitalizing Inner City Parks: New Funding Options Can Address the Needs of Underserved Urban Communities." Accessed November 19, 2019. https://www.nrpa.org/contentassets/f768428a39aa4035ae-55b2aaff372617/urban-parks.pdf

The Nature of Cities. 2014. "Do Urban Green Corridors 'Work'? It Depends on What We Want Them to Do. What Ecological and/or Social Function Can We Realistically Expect Green Corridors to Perform in Cities? What Attributes Define Them, from a Design and Performance Perspective?" Accessed August 30, 2019. https://www.thenatureofcities.com/2014/10/05/do-urban-green-corridors-work-it-depends-on-what-we-want-them-to-do-what-ecological-andor-social-functions-can-we-realis-tically-expect-green-corridors-to-perform-in-cities-what-attributes-defi/

Nelson, Arthur C., and Robert E. Lang. 2009. *The New Politics of Planning: How States and Local Governments Are Coming to Common Ground on Reshaping America's Built Environment*. Washington, DC: Urban Land Institute.

New World Encyclopedia. 2018. "Irrigation." Accessed October 7, 2019. https://www.newworldencyclopedia.org/entry/Irrigation

Nielson-Pincus, Max, Patricia Sussman, Drew Bennett, Hannah Gosnell, and Robert Parker. 2017. "The Influence of Place on the Willingness to Pay for Ecosystem Services." *Society & Natural Resources* 30(1): 1423–1441.

Nilon, Charles H., Myla F. J. Aronson, Sarel S. Cilliers, Cynnamon Dobbs, Lauren J. Frazee, Mark A. Goddard, Karen M. O'Neill, et al. 2017. "Planning for the Future of Urban Biodiversity: A Global Review of City Scale Initiatives." *Bioscience* 67(4): 332–342.

NOAA. 2018. "High Temperatures Bring Citizen Scientists to Map the Hottest Places in D.C. and Baltimore." Accessed August 18, 2019. https://research.noaa.gov/article/

ArtMID/587/ArticleID/2385/High-temperatures-bring-citizen-scientists-to-map-the-hottest-places-in-Baltimore-and-DC

NOAA Climate. n.d. "Climate Change and Citizen Science." Accessed September 1, 2019. https://research.noaa.gov/article/ArtMID/587/ArticleID/2385/mediaid/1389

Novel Ecosystems. 2016. "Novel Ecosystems." Accessed August 27, 2019. https://novel-ecosystems.russell.wisc.edu/novel-ecosystems-overview/

NPIC (National Pesticide Information Center). 1999. "DDT." Accessed December 30, 2019. http://npic.orst.edu/factsheets/ddtgen.pdf

NRCS USDA. 2000. "Urban Soil Compaction, Soil Quality: Urban Technical Note No. 2." Accessed July 19, 2019. https://www.nrcs.usda.gov/Internet/FSE_DOCUMENTS/nrcs142p2_053278.pdf

Nunno, Richard. 2018. "Fact Sheet: High Speed Rail Development Worldwide." Environmental and Energy Study Institute. Accessed December 3, 2019. https://www.eesi.org/papers/view/fact-sheet-high-speed-rail-development-worldwide

NYBG, New York Botanical Program. 2019. "New York City EcoFlora Project." Accessed July 19, 2019. https://www.nybg.org/science-project/new-york-city-ecoflora/

Oke, T. R. 2011. "Urban Heat Islands." In *The Routledge Handbook of Urban Ecology*, edited by Ian Douglas, David Goode, Mike Houck and Rusong Wang, 120–131. New York: Routledge.

Olmsted, Frederick Law. 1870. *Public Parks and the Enlargement of Towns*. Cambridge, MA: Riverside Press.

Ordeñana, Miguel. 2013. "L.A. Nature Map." Accessed November 16, 2019. http://www.inaturalist.org/projects/l-a-nature-map

The Oregon Conservation Strategy. n.d. "Conservation in Urban Areas." Accessed July 24, 2019. https://www.oregonconservationstrategy.org/conservation-toolbox/conservation-in-urban-areas/

Orff, Kate. 2016. *Toward an Urban Ecology*. New York: Monacelli Press.

Overview of the Millennium Ecosystem Assessment. n.d. Accessed July 29, 2019. https://millenniumassessment.org/en/About.html

Palmer, Matt. 2012. "Discovering Urban Biodiversity." Accessed August 5, 2019. https://www.thenatureofcities.com/2012/08/14/discovering-urban-biodiversity/

Panagos, Panos, Luca Montanarella, and Arwyn Jones. 2006. "Soil Related Policies in EU: the EU Thematic Strategy on Soil Protection." Accessed July 10, 2020 https://archive.epa.gov/oswer/international/web/html/200906_eu_soils_policy.html

Parlow, Eberhard. 2011. "Urban Climate." In *Urban Ecology: Patterns, Processes, and Applications*, edited by Jari Niemelä, Jurgen Breuste, Thomas Elmqvist, Glenn Guntenspergen, Philip James, and Nancy E. McIntyre, 31–43. New York: Oxford University Press.

Pennisi, Elizabeth. 1991. "Researchers Work to Give Biodiversity a Scientific Identity." Accessed August 7, 2019. https://www.the-scientist.com/research/researchers-work-to-give-biodiversity-a-scientific-identity-60657

Peters, Adele. 2019. "Cities are Getting Hotter, but We Can Redesign Them to Keep Us Cool." Accessed August 16, 2019. https://fastcompany.com/90379081/cities-are-getting-hotter

Peterson, M. Nils, Brandi Thurmond, Melissa McHale, Shari Rodriguez, Howard D.

Bondell, and Merril Cook. 2012. "Predicting Native Plant Landscaping Preferences in Urban Areas." *Sustainable Cities and Society* 5: 70–76.

Petri, Alexandra. 2017. "10 Easy Ways You Can Help Scientists Study the Earth." Accessed September 1, 2019. https://www.nationalgeographic.com/news/2017/04/citizen-science-projects-environment-climate-change-weather/#close

Piacentini, Richard. 2019. "Center for Sustainable Landscapes Achieves SITES Platinum Certification." Accessed October 3, 2019. https://thefield.asla.org/2019/04/30/center-for-sustainable-landscapes-sites-platinum-certification/

Pickett, Mallory. 2017. "Around the Pier: Scripps Scientists Help Create and Distribute a 'Smartfin' that turns Surfers into Citizen Scientists." Accessed September 2, 2019. https://scripps.ucsd.edu/news/around-pier-scripps-scientists-help-create-and-distribute-smartfin-turns-surfers-citizen

Pickett, Steward T. A., Geoffrey L. Buckley, Sujay S. Kaushal, and Yvette Williams. 2011. "Social-ecological Science in the Humane Metropolis." *Urban Ecosystems* 14(3): 319–339.

Pickett, Steward T. A., Mary L. Cadenasso, Daniel L. Childers, Mark J. McDonnell, and Weiqi Zhou. 2016. "Evolution and Future of Urban Ecological Science: Ecology *in, of,* and *for* the City." *Ecosystem Health and Sustainability* 2(7): e01229. doi:10.1002/ehs2.1229

Plastrik, Pete. 2018. "Can It Happen Here? Managed Retreat for US Cities." Accessed August 30, 2019. http://lifeaftercarbon.net/2018/08/can-it-happen-here-managed-retreat-for-us-cities/

Pollan, Michael. 1989. "Why Mow? The Case Against Lawns." Accessed November 12, 2019. https://michaelpollan.com/articles-archive/why-mow-the-case-against-lawns/

Polycarpou, Lakis. 2010. "The Problem of Lawns." *State of the Planet, Earth Institute, Columbia University*. Accessed September 29, 2019. https://blogs.ei.columbia.edu/2010/06/04/the-problem-of-lawns/

Pooley, Julie Ann, and Moira O'Conner. 2000. "Environmental Education and Attitudes: Emotions and Beliefs are What is Needed." *Environment and Behavior* 32(September): 711–723.

Poon, Linda. 2018. "Every Tree in the City, Mapped." Accessed July 17, 2019. https://www.citylab.com/environment/2018/12/urban-tree-canopy-maps-artificial-intelligence-descartes-labs/578701/

Postel, Sandra. 2017. *Replenish: The Virtuous Cycle of Water and Prosperity*. Washington, DC: Island Press.

Pritzlaff, Richard. 2019. "Ecological, Conservation & Ecosystem Services Terminology: a short glossary of words and terms to understand." Accessed July 28, 2019. https://www.biophiliafoundation.org/ecological-terminology/

Project for Public Spaces. 2007. "What Is Placemaking?" Project for Public Spaces. Accessed October 27, 2019. https://www.pps.org/article/what-is-placemaking

Puppim de Oliveira, Jose, Christopher Doll, Raquel Moreno-Penaranda, and Osman Balaban. 2014. "Urban Biodiversity and Climate Change." In *Global Environmental Change, Handbook of Global Environmental Pollution*, vol 1, edited by B. Freedman, 461–468. Dordrecht: Springer.

Quigley, Martin. 2011. "Chapter 2.2: Potemkin Gardens: Biodiversity in Small Designed

Landscapes." In *Urban Ecology: Patterns, Processes, and Applications*, edited by Jari Niemelä, Jurgen Breuste, Thomas Elmqvist, Glenn Guntenspergen, Philip James, and Nancy E. McIntyre, 85–92. New York: Oxford University Press.

Raddick, Jordan M., Georgia Bracey, Pamela L. Gay, Chris J. Lintott, Phil Murray, Kevin Schawinski, Alexander S. Szalay, and Jan Vandenberg. 2010. "Galaxy Zoo: Exploring the Motivations of Citizen Science Volunteers." *Astronomy Education Review* 9: 910.

Radford, Gail. 1996. *Modern Housing for America: Policy Struggles in the New Deal Era.* Chicago: University of Chicago Press.

Richards, Daniel R., and Benjamin S. Thompson. 2019. "Urban Ecosystems: A New Frontier for Payments for Ecosystem Services." *People and Nature* 1(2): 1–34.

Rigolon, Alessandro. 2016. "A Complex Landscape of Inequity in Access to Urban Parks: A Literature Review." *Landscape and Urban Planning* 153: 160–169.

Rigolon, Alessandro, Matthew Browning, and Viniece Jennings. 2018. "Inequities in the Quality of Urban Park Systems: An Environmental Justice Investigation of Cities in the United States." *Landscape and Urban Planning* 178: 156–169.

RinkWatch. n.d. "RinkWatch: Where Skating Meets Environmental Science." Accessed May 4, 2021. https://www.rinkwatch.org/

Rittel, Horst W. J., and Melvin M. Webber. 1973. "Dilemmas in a General Theory of Planning." *Policy Sciences* 4(2): 155–169.

Ritter, Jim. 2018. "Soil Erosion—Causes and Effects." Accessed July 11, 2019. www.omafra.gov.on.ca/english/engineer/facts/12-053.htm.

Robbins, Paul, and Trevor Birkenholtz. 2003. "Turfgrass Revolution: Measuring the Expansion of the American Lawn." *Land Use Policy* 20: 181–194.

Robinson, Judith Helm, Noel D. Vernon, and Catherine C. Lavoie. 2005. "Historic American Landscape Survey Guidelines for Historical Reports." U.S. Department of the Interior, National Park Service, Historic American Buildings Survey/Historic American Engineering Record/Historic American Landscapes Survey. Accessed October 10, 2019. https://www.nps.gov/hdp/standards/HALS/HALSHistoryGuidelines.pdf

Rose, Nancy Ellen. 2009. *Put to Work: The WPA and Public Employment in the Great Depression.* 2nd ed. New York: Monthly Review Press. Original edition, 1994.

Rosenberg, David, E. Kelly Kopp, Heidi A. Kratsch, Larry Rupp, Paul Johnson, and Roger Kjelgren. 2011. "Value Landscape Engineering: Identifying Costs, Water Use, Labor, and Impacts Supporting Landscape Choice." *Journal of The American Water Resources Association* 47(3): 635–649.

Rosenzweig, C., W. Solecki, P. Romero-Lankao, S. Mehrotra, S. Dhakal, T. Bowman, and S. Ali Ibrahim. 2015. "ARC3.2 Summary for City Leaders. Climate Change and Cities, Second Assessment Report of the Urban Climate Change Research Network." Urban Climate Change Research Network. Columbia University, New York. Accessed August 31, 2019. http://uccrn.org/files/2015/12/ARC3-2-web.pdf

Rosiak, Luke. 2013. "Exography: LEED Certification Doesn't Guarantee Energy Efficiency, Analysis Shows." Accessed October 1, 2019. https://www.washingtonexaminer.com/exography-leed-certification-doesnt-guarantee-energy-efficiency-analysis-shows

Rybczynski, Witold. 1995. *City Life: Urban Expectations in a New World.* New York: Scribner.

Sadeghian, Mohammad Mehdi, and Zhirayr Vardanyan. 2015. "A Brief Review on Urban Park History, Classification and Function." *International Journal of Scientific & Technology Research* 4(11): 120–124.

SAIC (Scientific Applications International Corporation). 2006. "Life Cycle Assessment: Principles and Practice." EPA/600/R-06/060.

Salk Institute. 2019. "Harnessing Plants Initiative: The Salk Approach." Accessed July 19, 2019. https://www.salk.edu/harnessing-plants-initiative/

Sauerwein, Martin. 2011. "Urban Soils—Characterization, Pollution, and Relevance in Urban Ecosystems." In *Urban Ecology, Patterns, Processes, and Applications*, edited by Jari Niemelä, Jurgen Breuste, Thomas Elmqvist, Glenn Guntenspergen, Philip James, and Nancy E. McIntyre, 103–114. New York: Oxford University Press.

Saunders, William S., and Kongjian Yu. 2012. *Designed Ecologies: The Landscape Architecture of Kongjian Yu*. Basel: Birkhäuser.

Schaefer, Valentin. 2017. "Incorporating Novel Ecosystems and Layered Landscapes for Ecological Restoration in Cities." *American Journal of Life Sciences* 5(6): 164–169.

Schlosberg, David. 2004. "Reconceiving Environmental Justice: Global Movements and Political Theories." *Environmental Politics* 13(3): 517–540.

Schmidt, Charles W. 2004. "Sprawl: The New Manifest Destiny?" *Environmental Health Perspectives* 112(11): A620–A627.

Schnaars, Christopher, and Hannah Morgan. 2012. "In U.S. Building Industry, Is It Too Easy to be Green?" *USA Today*, October 24, 2012. Accessed October 1, 2019. https://www.usatoday.com/story/news/nation/2012/10/24/green-building-leed-certification/1650517/

Schön, Donald A. 1983. *The Reflective Practitioner: How Professionals Think in Action*. New York: Basic Books.

Schröter, Matthias, Roland Kraemer, Martin Mantel, Nadja Kabisch, Susanne Hecker, Anett Richter, Veronika Neumeier, and Aletta Bonn. 2017. "Citizen Science for Assessing Ecosystem Services: Status, Challenges and Opportunities." *Ecosystem Services* 28: 80–94.

Schwartz, Lisa. 2007. "An Interview with Ecologist Mike Rosenzweig: Exploring Reconciliation Ecology." Accessed December 8, 2019. http://tolweb.org/tree/ToLmovies/rosenzweiginterview1.mp4

SciStarter. 2017. "City Nature Challenge 2017: The Wasatch Front." Accessed March 10, 2021. https://scistarter.org/city-nature-challenge-2017-the-wasatch-front-1

———. 2019a. "What is a Citizen Scientist?" Accessed November 30, 2019. https://scistarter.org/citizen-science

———. 2019b. "Canberra Nature Map." Accessed January 12, 2020. https://scistarter.org/canberra-nature-map

———. 2019c. "Catch the Tide Miami—Peeking at Future Sea Levels." Accessed November 19, 2019. https://scistarter.org/catch-the-tide-miami-peeking-at-future-sea-levels

Scott, Tommi Jo, Alyssa Politte, Sean Saathoff, Sam Collard, Emily Berglund, Joshua Barbour, and Alex Sprintson. 2014. "An evaluation of the Stormwater Footprint Calculator and the Hydrological Footprint Residence for Communicating about Sustainability in Stormwater Management." *Sustainability: Science, Practice, & Policy* 10(2): 51–64.

Seiter, David. 2011. "Profiles of Spontaneous Plants." Accessed July 17, 2019. https://urbanomnibus.net/2011/12/profiles-of-spontaneous-urban-plants/

———. 2015. "Spontaneous Urban Plants." Accessed July 17, 2019. https://www.asla.org/2015awards/96909.html

Sidwell (Sidwell Friends School). 2019. Accessed November 22, 2019. https://www.sidwell.edu/about/environmental-stewardship

Simpson, Greg, and David Newsome. 2017. "Environmental History of an Urban Wetland: From Degraded Colonial Resource to Nature Conservation Area." *Geography and Environment* 4(1): 1–18

Sisser, John M., Kristen C. Nelson, Kelli L. Larson, Laura A. Ogden, Colin Polsky, and Rinku Roy Chowdhury. 2016. "Lawn Enforcement: How Municipal Policies and Neighborhood Norms Influence Homeowner Residential Landscape Management." *Landscape and Urban Planning* 150: 16–25.

SITES. 2009. "The Sustainable Sites Initiative: The Case for Sustainable Landscapes." American Society of Landscape Architects, Lady Bird Johnson Wildflower Center at the University of Texas at Austin, and United States Botanic Garden. https://www.usgbc.org/resources/sites-rating-system-and-scorecard

———. 2014. "SITES v2 Rating System for Sustainable Land Design and Development." Accessed October 1, 2019. https://www.asla.org/uploadedFiles/CMS/AboutJoin/Copy%20of%20SITESv2_Scorecard%20Summary.pdf

Smart Growth. n.d. "Smart Growth Principles." Accessed April 18, 2021. https://smartgrowth.org/smart-growth-principles/

Solly, Meilan. 2019. "California Will Build the Largest Wildlife Crossing in the World." Accessed November 12, 2019. https://www.smithsonianmag.com/smart-news/california-will-build-largest-wildlife-crossing-world-180972947/

State of Victoria. 2002. "Melbourne 2030: Planning for Sustainable Growth." Melbourne: Victoria Department of Infrastructure. Accessed November 28, 2019. https://www.planning.vic.gov.au/__data/assets/pdf_file/0022/107419/Melbourne-2030-Full-Report.pdf

Stepenuck, Kristine F., and Linda T. Green. 2015. "Individual and Community-Level Impacts of Volunteer Environmental Monitoring: A Synthesis of Peer-Reviewed Literature." *Ecology and Society* 20(3): 8–23.

The Stormwater Report. 2014. "First Full-Scale Water Square Opens in Rotterdam." Accessed August 30, 2019. https://stormwater.wef.org/2014/03/first-full-scale-water-square-opens-rotterdam/

Study.com. 2019. "Natural Resource Manager Career Summary." Accessed November 19, 2019. https://study.com/articles/Natural_Resource_Manager_Career_Summary.html

———. 2020. "What is Environmental Studies: Definition and Topics." Accessed November 19, 2019. https://study.com/academy/lesson/what-is-environmental-studies-definition-topics.html

Szlavecz, Katalin. n.d. "About GLUSEEN." Accessed August 6, 2019. http://www.gluseen.org

Tachieva, Galina. 2010. *Sprawl Repair Manual*. Washington, DC: Island Press.

Takada, Aya. 2018. "As High-Rise Farms Go Global, Japan's Spread Leads the Way." Accessed December 4, 2019. https://www.japantimes.co.jp/news/2018/11/01/business/tech/high-rise-farms-go-global-japans-spread-leads-way/#.XiJq4S2ZOF0

Thompson, Luke. 2019. "The Earth Microbiome Project is a Systematic Attempt to Characterize Global Microbial Taxonomic and Functional Diversity for the Benefit of the Planet and Humankind." Accessed July 11, 2019. www.earthmicrobiome.org.

TPL (The Trust for Public Land). 2016. "City Parks, Clean Water: Making Great Places Using Green Infrastructure." San Francisco: The Trust for Public Land.

The Trust for Public Land. 2016. "City Parks, Clean Water. Making Great Places Using Green Infrastructure." Accessed October 8, 2019. https://www.tpl.org/sites/default/files/City%20Parks%20Clean%20Water%20report_0.pdf

TRY. 2019. "TRY Plant Trait Database." Accessed August 8, 2019. https://www.try-db.org/

Tuhus-Dubrow, Rebecca. 2014. "Pretty Park, Affordable Rent: Making Neighborhoods 'Just Green Enough.'" Accessed August 27, 2019. https://nextcity.org/daily/entry/gentrification-green-neighborhoods-just-green-enough

Tveit, M., A. Ode, and G. Fry. 2006. "Key Concepts in a Framework for Analyzing Visual Landscape Character." *Landscape Research* 31(3): 229–255.

UC IPM. 2019. "What Is Integrated Pest Management (IPM)?" University of California Agriculture and Natural Resources Statewide Integrated Pest Management Program. Accessed November 5, 2019. https://www2.ipm.ucanr.edu/What-is-IPM/

Ueda, Ken-ichi. 2018. "City Nature Challenge 2019." Accessed July 23, 2019. https://www.inaturalist.org/projects

UF/IFAS CAIP. 2018. "Non-native Invasive Plants—An Introduction." Center for Aquatic and Invasive Plants, University of Florida. Accessed November 4, 2019. http://plants.ifas.ufl.edu/manage/why-manage-plants/non-native-invasive-plants-an-introduction/

Ulrich, Roger S. 1984. "View through a Window May Influence Recovery from Surgery." *Science* 224(4647): 420–421.

Ulrich, Roger S., Robert F. Simons, Barbara D. Losito, Evelyn Fiorito, Mark A. Miles, and Michael Zelson. 1991. "Stress Recovery during Exposure to Natural and Urban Environments." *Journal of Environmental Psychology* 11(3): 201–230.

UN (United Nations). 1973. *Report of the United Nations Conference on the Human Environment*. New York: United Nations.

———. 1987. *Our Common Future: Report of the World Commission on Environment and Development*, edited by World Commission on Environment and Development (WCED). Oxford: Oxford University Press.

———. 2019a. "What is the Kyoto Protocol?" Accessed December 4, 2019. https://unfccc.int/kyoto_protocol

———. 2019b. "The Paris Agreement." Accessed December 4, 2019. https://unfccc.int/process-and-meetings/the-paris-agreement/the-paris-agreement

———. 2019c. "The Untapped Potential of Citizen Science to Track Progress on the Sustainable Development Goals." *UN Environment*, October 11, 2019. Accessed December 30, 2019. https://www.unenvironment.org/news-and-stories/story/untapped-potential-citizen-science-track-progress-sustainable-development

UNCED (United Nations Conference on Environment and Development). 1993. *Agenda 21, Programme of Action for Sustainable Development: Rio Declaration on Environ-*

ment and Development. New York: United Nations Department of Public Information.

UN Environment Programme. n.d. "Convention of Biological Diversity." Accessed December 11, 2019. https://www.un.org/en/events/biodiversityday/convention.shtml

———. 2017. "Indigenous People and Nature: A Tradition of Conservation." Accessed November 11, 2019. https://www.unenvironment.org/news-and-stories/story/indigenous-people-and-nature-tradition-conservation

UNESCO Digital Library. 2004. "Urban Biosphere Reserves in the Context of the Statutory Framework and the Seville Strategy for the World Network of Biosphere Reserves." Accessed November 17, 2019. https://unesdoc.unesco.org/ark:/48223/pf0000 136414?posInSet=1&queryId=10d6e1ea-d344–4b12-b6fe-6ce1b0780c84

UNESCO. 2011. "Recommendation on the Historic Urban Landscape." Accessed October 13, 2019. https://whc.unesco.org/uploads/activities/documents/activity-638-98.pdf

———. 2013. "New Life for Historic Cities: The Historic Urban Landscape Approach Explained." Accessed October 9, 2019. https://unesdoc.unesco.org/ark:/48223/pf0000220957

———. 2017. "A New Roadmap for the Man and the Biosphere (MAB) Programme and its World Network of Biosphere Reserves." United Nations Educational, Scientific and Cultural Organization. Paris: UNESCO. https://unesdoc.unesco.org/ark:/48223/pf0000247418

United States Botanic Garden. 2014. "Supporting the Health of Honey Bees and Other Pollinators." Addendum to *Guidance for Federal Agencies on Sustainable Practices for Designed Landscapes.* 2010. Accessed November 7, 2019. https://www.fws.gov/southwest/es/Documents/R2ES/Pollinators/6-Supporting_the_Health_of_Honey_Bees_and_Other_Pollinators_Oct2014.pdf

United States Department of Agriculture. n.d. "Soil Health Nuggets: There are Some Amazing Things Going on Underground." Accessed July 17, 2019. https://www.nrcs.usda.gov/Internet/FSE_DOCUMENTS/stelprdb1101660.pdf

United States Government. 2019. "Federal Laws and Regulations." Accessed December 13, 2019. https://www.usa.gov/laws-and-regs

University of Oklahoma. 2018. "What's in Your Back Yard? The University of Oklahoma Citizen Science Soil Collection Program." Accessed July 16, 2019. https://whatsinyourbackyard.org

Urban Ecology Field Station. 2019. "What is Urban Ecology?" Indiana University Southeast. Accessed November 19, 2019. https://www.ius.edu/field-station/what-is-urban-ecology.php

The Urban Wildlife Working Group. 2012. "Disturbances and Threats to Urban Wildlife." Accessed July 24, 2019. https://urbanwildlifegroup.org/urban-wildlife-information

Uren, Hannah V., Peta L. Dzidic, Brian J. Bishop. 2015. "Exploring Social and Cultural Norms to Promote Ecologically Sensitive Residential Garden Design." *Landscape and Urban Planning* 137: 76–84.

USEPA GreenScape Tools. 2016. "Resource Conserving Landscape Cost Calculator." Accessed October 14, 2019. https://www.landscapeperformance.org/benefits-toolkit/resource-conserving-landscaping-cost-calculator

USEPA (United States Environmental Protection Agency). 2008a. "Urban Heat Island Basics." In *Reducing Urban Heat Islands: Compendium of Strategies*. Accessed August 17, 2019. https://www.epa.gov/sites/production/files/2017-05/documents/reducing_urban_heat_islands_ch_1.pdf

———. 2008b. "Heat Island Reduction Activities." In *Reducing Urban Heat Islands: Compendium of Strategies*. Accessed August 17, 2019. https://www.epa.gov/sites/production/files/2017-05/documents/reducing_urban_heat_islands_ch_6.pdf

———. 2008c. "Cool Roofs." In *Reducing Urban Heat Islands: Compendium of Strategies*. Accessed August 17, 2019. https://www.epa.gov/sites/production/files/2017–05/documents/reducing_urban_heat_islands_ch_4.pdf

———. 2008d. "Cool Pavements." In *Reducing Urban Heat Islands: Compendium of strategies*. Accessed August 17, 2019. https://www.epa.gov/sites/production/files/2017–05/documents/reducing_urban_heat_islands_ch_5.pdf

USGBC. 2016. "Stakeholder Feedback Report. Built by and for Cities and Communities. LEED v4.1 Cities and Communities: Existing." Accessed October 1, 2019. https://www.usgbc.org/resources/leed-cities-stakeholder-feedback-report-july-2019

Utah Department of Environmental Quality. n.d. "Construction and Demolition Pollution Prevention Fact Sheet." Accessed July 8, 2019. https://digitallibrary.utah.gov/awweb/awarchive?type+download&item=16569

Utrecht University. 2016. "Can Drinking Tea Help Us Understand Climate Change?" http://www.teatime4science.org/about/the-project/

Villanova University. 2019. "City Manager Job Description." Accessed November 19, 2019. https://www.villanovau.com/resources/public-administration/public-administration-city-manager-job-description/

Wandersee, James, and Elisabeth Schussler. 1999. "Preventing Plant Blindness." *American Biology Teacher* 61(2): 82–86.

Ward Thompson, Catharine. 2011. "Linking Landscape and Health: The Recurring Theme." *Landscape and Urban Planning* 99(3–4): 187–195.

Warner, Keith Douglass, and David DeCosse. 2009a. "Thinking Ethically about the Environment." *A Short Course in Environmental Ethics: Introduction*. Accessed November 10, 2019. https://www.scu.edu/environmental-ethics/short-course-in-environmental-ethics/

———. 2009b. "An Autobiography of Your Relationship with the Earth." *A Short Course in Environmental Ethics: Lesson One*. Accessed November 10, 2019. https://www.scu.edu/environmental-ethics/short-course-in-environmental-ethics/lesson-one/

———. 2009c. "Who, When, Where, and How: The Distinctiveness of Environmental Ethics." *A Short Course in Environmental Ethics: Lesson Two*. Accessed November 10, 2019. https://www.scu.edu/environmental-ethics/short-course-in-environmental-ethics/lesson-two/

———. 2009d. "The Role of Science in Environmental Ethics." *A Short Course in Environmental Ethics: Lesson Seven*. Accessed November 10, 2019. https://www.scu.edu/environmental-ethics/short-course-in-environmental-ethics/lesson-seven/

Warner, Lara A., Amanda D. Ali, and Anil Kumar Chaudhary. 2017. "Residents' Perceived Landscape Benefits Can Help Extension Promote Good Landscape Management Practices." AEC620, Department of Agricultural Education and Communication, UF/IFAS Extension. https://edis.ifas.ufl.edu/wc282

Warner, Lara A., and Paul Monaghan. 2016. "Using Audience Commitment to Increase Behavior Changes in Sustainable Landscaping." WC154, Department of Agricultural Education and Communication, UF/IFAS Extension. https://edis.ifas.ufl.edu/wc154

WAV. 2007. "Water Action Volunteers Stream Monitoring Program." Accessed January 19, 2020. http://watermonitoring.uwex.edu/wav/

Webb, R. 2009. "Dockside Green: A Model for Creating Affordable Housing and Inclusionary Communities." *Canadian Institute of Planners Conference*. University of Manitoba.

Weinstein, Alan C. 2005. "Homeowners Associations." *Planning Commissioners Journal* 58: 1–6.

Whiteman, Lily. 2012. "YardMap Helps Wildlife in Your Backyard, Literally." Accessed October 22, 2019. https://www.livescience.com/22271-yardmap-citizen-science-conservation-nsf-bts.html

WHO (World Health Organization). n.d. "Ambient Air Pollution: Health Impacts—World Health Organization." Accessed August 15, 2019. https://www.who.int/airpollution/ambient/health-impacts/en/

———. 2019. *World Malaria Report 2019*. https://www.who.int/malaria/publications/world-malaria-report-2019/en/

Wilen, Cheryl A. 2018. "Weed Management in Landscapes." Pest Notes, Publ. 7441, UC IPM. Accessed July 16, 2019. http://ipm.ucanr.edu/PMG/PESTNOTES/pn7441.html

Wilson, Edward O. 1984. *Biophilia*. Cambridge, MA: Harvard University Press.

Wolch, Jennifer R., Jason Byrne, and Joshua P. Newell. 2014. "Urban Green Space, Public Health, and Environmental Justice: The Challenge of Making Cities 'Just Green Enough.'" *Landscape and Urban Planning* 125: 234–244.

Wolf, K. L., S. Krueger, and K. Flora. 2014. "Place Attachment and Meaning—A Literature Review." *Green Cities: Good Health*. Accessed November 1, 2019. https://depts.washington.edu/hhwb/Thm_Place.html

Worrall, Simon. 2017. "Without Bugs We Might All Be Dead." *National Geographic*, August 5, 2017. Accessed July 25, 2019. https://www.nationalgeographic.com/news/2017/08/insect-bug-medicine-food-macneal/

Yang, Bo, Shujuan Li, and Chris Binder. 2016. "A Research Frontier in Landscape Architecture: Landscape Performance and Assessment of Social Benefits." *Landscape Research* 41(3): 314–329.

Young, Dean, and Chris Morrison. 2012. "Soil Management Best Practices Guide for Urban Construction." Accessed July 8, 2019. https://trieca.com/app/uploads/2016/07/Soil-Mgmt-Guideline-Mar-28-2012.pdf

Zevenbergen, Chris, Dafang Fu, and Assela Pathirana. 2018. "Transitioning to Sponge Cities: Challenges and Opportunities to Address Urban Water Problems in China." *Water* 10(9): 1230.

Index

Page numbers in *italics* indicate illustrations.

industrial districts and, 127–28; as ecological traps, 82–83; embedding cities in parks and bioreserves, 123–24; environmental accreditation system for, 134; environmental justice and, 108, 247, 249, 251–52; features of, 121; generating support for, 124–25, 128; golf courses as, 95, 121, 176; historic influences on, 151, 163, 165; innovative strategies for, 126–28; intentional/unintentional, 95; IPM and IVM practices, 173–74; management of, 105; mental health benefits and, 90, 152, 204; mitigation of human impact on, 170–73; non-native species and, 172; novel ecosystems and, 95, 100, 103, 105, 107; parklets and, 126–27, 127; personal attachment to, 1, 197, 199–200, 202; planning and design of, 82; as population sinks, 82–83; private developments and, 121–22; recreation and, 95, 120, 151; residential landscapes as, 153, 170, 283; restoration of, 163, 165; social reform movement and, 160; soils and, 19, 21, 23; strategies for creating, 124–25; temperature and, 68, 71; temporary mobile, 126; turfgrass and, 92; urban biodiversity and, 3, 76, 79, 82–83; urban nature and, 158; urban vegetation and, 44, 49, 74, 76; urban wildlife and, 52, 58, 121; users and purposes of, 121–23; value of, 82, 93, 227; vision for, 120; wealthy and, 234; well-being and, 90–91, 120, 197, 204

Green Star, 134
Green urbanism, 142–44
Green walls, 49, 70, 71, 130, 131, 197
Greenways for Wildlife Project, 58
Groundwork RVA, 72

Habitat destruction, 78–79
Habitat restoration: citizen involvement in, 59; conservation planning and, 59, 82; green infrastructure and, 113–14; novel ecosystems and, 99–101, 105–6; remnant vegetation and, 81; urban soils and, 19–20; wetland and stream, 29–30, 36, 111, 113
Halifax, Canada, 140
Halprin, Lawrence, 132
Hampton Court Palace, 160
Harbin, China, 146, 147, 219, 305, 308, 313
Harbor seals, 58
Hardscape materials, 130–31
Harnessing Plants Initiative (HPI), 49

Health: air pollution and, 186; drug research and, 19–20, 27–28; environmental hazards and, 247; environmental justice and, 153; glyphosate and, 48; green infrastructure and, 153; heat stress and, 66–67; neighborhoods and, 153, 255, 260–61; obesity and, 260; PAHs and, 24; soil contamination and, 19, 24; suburban development and, 260; urban heat islands and, 18, 66–67; urban hydrology and, 29–30, 34; urbanization and, 257; urban soils and, 17, 19–20; urban wildlife and, 56; water pollution and, 186
Heat budgets, 65–67
Heat convection, 66
Heat fluxes, 65–66
Heat islands. See Urban heat islands (UHIs)
Heat maps, 63, 68, 72–73
Heat-proofing, 143–44
Heat stress, 66–67
Heavy metals, 24, 34
Herbicides, 47, 174, 246, 272
Hercules, California, 209
Heritage sites, 231
High Line Park (New York City), 125, 127–28
High-speed rails (HSR), 311–12
Historic American Landscapes Survey (HALS), 163
Historic landscapes, 163–65
Homeowner associations (HOA): certification program conflicts, 301–2; Covenants, Conditions, and Restrictions (CC&Rs), 181; cultural norms and, 222; ecological aesthetics and, 272; environmental ethics and, 222; expansion of, 183; governance by, 180–83; infringement on private rights by, 181, 183; landscape codes, 41, 47, 211, 222, 276, 301; lawn restrictions and, 211, 222, 271, 301–2; weeds and, 47
HPI. See Harnessing Plants Initiative (HPI)
Humans: ecological identity and, 202; ecosystems and, 18, 88; heat budgets and, 66–67; impact on biological systems, 4–5, 17; mitigating impact of, 168–73; nature and, 1–3, 57, 85–86, 90–91, 187–88, 192, 223; plant blindness and, 50–51; shared spaces for wildlife, 58; spread of invasive plants and, 172–73; urban nature and, 103, 105; wildlife encounters, 56–60

low-impact development (LID) and, 137; rulemaking in, 179; stormwater management and, 30, 111, 116; urban ecology studies in, 154; zoning laws in, 180

United States Botanic Garden, 297

Universal Floristic Quality Assessment (FAQ) Calculator, 243

University of Florida, 13, 32, 101, 299

University of Oklahoma Citizen Science Soil Collection Program, 20, 27–28

Urban biodiversity: aesthetic appeal and, 77–78; avian, 76, 81; bird databases, 81; citizen science projects and, 83–84, 195; conservation biology and, 82–83; creation of, 80–83; data collection on, 75–76, 80–82; defining, 76; economic value and, 75, 79; ecosystem services and, 79–80, 107–8; green spaces and, 82–83; human-designed landscapes, 76–78; human well-being and, 75; importance of, 78–79; native birds and, 76; native plants and, 23, 74, 78, 81, 105; natural biodiversity versus, 76–77; non-native species and, 76, 80, 106; novel ecosystems and, 76–77; plant databases, 81; plant species variety and, 76; research in, 74–77, 155; resilience and, 78; social dimensions of, 75, 77–78; wildlife and, 52. See also Biodiversity

Urban Biodiversity Research Coordination Network (UrBioNet), 81

Urban Biosphere Reserve (UBR), 123

Urban boundary layer (UBL), 65

Urban canopy layer (UCL), 64

Urban canyons, 64, 64, 66

Urban climate: air pollution and, 62; characteristics of, 63–66; citizen science projects and, 63, 72–73; cold cities and, 67; cool construction materials for, 69–70; health impacts of, 66–67; heat budgets and, 65, 67; heat islands and, 4, 18, 62–73; humidity and rainfall, 66; nighttime cooling, 65–66; paved areas and, 68–69; street canyons in, 64, 64, 66; thermal water pollution and, 67; wind movement and, 66

Urban development: adaptation to environmental pressures, 141; biodiversity and, 74; biophilic design and, 189–90; ecosystem services and, 93; environmental ethics

and, 223–25; green infrastructure and, 111–17; historic influences on, 151; impact on natural environment, 168; impact on water, 111; international aid agencies and, 183–84; natural resource strategies, 151–52; novel ecosystems and, 76

Urban ecology: applied, 7–8; applied sciences and, 287–88; certification programs and, 296–98; citizen participation in, 11, 291–93; college programs and, 291–92; cultural aspects of, 11; culture-based knowledge, 286; decision-makers in, 15; defining, 2–3; development of, 5–6; environmental ethics in, 217–19; future cities and, 304–15; global networks for, 292; historic influences on, 157–58, 163–64; human dimension of, 9–11; interconnected systems in, 1–3; meaning and emotion in, 1, 204; paradigm shift in, 106–7; place attachment and, 197, 202; professions related to, 288–91; research in, 154–55; science-based knowledge, 286–88; science of cities and, 8–9; social marketing programs, 154; social sciences and, 59

Urban ecosystems: biodiversity in, 17, 23, 74, 76–77; bionetworks and, 95–96; climate change and, 144, 146, 149; defining, 18; diversifying functions, 107; ecological gentrification and, 108; ecosystem approach and, 97; humans in, 85; IPM and IVM practices, 173; layered landscapes in, 105; non-native species and, 107; novel, 76–78; socio-ecological systems and, 97

Urban ecosystem services: biodiversity in, 107–8; citizen science projects and, 86, 93–94; cultural benefits, 90–92; defining, 86; environmental policy and, 85; green and blue (water) areas, 89; importance of, 89–92; investment strategies, 92–93; market economics and, 90, 92–93; population benefits and, 89; provisioning, 90; regulating activities, 90; social capital and, 86; supporting functions, 90; sustainability and, 92; value of, 86, 89–90, 93, 227–28, 233–34. See also Ecosystem services (ES)

Urban farms, 25, 52, 80, 310

Urban green spaces. See Green spaces

Urban heat budgets, 65–67

Gail Hansen is associate professor in the Environmental Horticulture Department of the University of Florida, Institute of Food and Agricultural Sciences. Prior to joining the university, Gail worked for a landscape architecture/land planning firm for eight years on various planning and urban design projects. Her areas of expertise in teaching and extension include design of future urban landscapes, urban ecology, and the social and cultural dimensions of urban landscapes. As the statewide extension specialist in urban landscape design, Gail has authored more than 30 extension publications and numerous teaching and design manuals on sustainable urban landscapes. She has also been an invited lecturer at universities in China, Australia, and Turkey, and her publications are used for teaching at universities in South Africa, Spain, Malaysia, Brazil, and Hungary. In her extension program she regularly consults with developers, city agencies, and community organizations on best design practices for sustainable urban landscapes. Dr. Hansen has an MLA and a PhD in landscape architecture from the University of Florida.

Hansen is the primary author of chapters 1–10, 14–16, 19–23, 26, 27, 29, and 30. Hansen and Macedo coauthored chapters 12 and 28.

*   *   *

Joseli Macedo is professor in the School of Architecture, Planning and Landscape at the University of Calgary and former dean of the Faculty of Architecture and Planning at Dalhousie University in Canada. Prior to moving to Canada, she served as Head of the School of Design and the Built Environment at Curtin University in Perth, Australia. She started her academic career at the University of Florida (UF) in the United States where she was Chair of the Department of Urban and Regional Planning, Director of the Center for International Design and Planning, and Affiliate Faculty in the Center for Latin American Studies and the School of Natural Resources and the Environment. During her tenure at UF, Joseli was awarded a Fulbright-Nehru Fellowship and lived and worked in Ahmedabad, India. She has also lived and worked in Brazil. Joseli has taught and conducted research in the areas of sustainable cities, urban design, and international development for 30 years. She has published on city design and urban form, land policy and land tenure, housing policy, urban planning history, and pedagogy.

Macedo is the primary author of chapters 11, 13, 17, 18, 24, 25.